T0139882

Primed for Success: The Story of Scientific Design Company

"More than a company history, *Primed for Success* is the story of the chemical industry in the United States. It is comprehensive in scope and detailed in its treatment—an essential read for anyone who studies the chemical industry or has been part of it."

—Thomas M. Connelly, *Chief Executive Officer, ACS*

"Peter Spitz' detailed and engrossing account of the rise of the petrochemical industry—which set in motion the building blocks of the modern industrial economy—captures how transformative the industry has been to not just economic development, but to so many aspects of modern life."

—Andrew N. Liveris, *Former Chairman and CEO, Dow Chemical*

"Peter Spitz is opening up a new chapter in the history of the petrochemical industry that provides a lasting platform and remembrance of the great entrepreneurs of the industry. The new book should be very valuable for the new generation of chemical engineers as they recognize the vision and creativity of many courageous scientists and engineers."

—Werner Praetorius, *President Petrochemicals (ret.), BASF Group*

"Peter Spitz delivers an essential history with *Primed for Success*. He details the inside story of a disruptive startup that invented the breakthrough technologies that would help usher in the petrochemical age. The book's insights on entrepreneurship, successful innovation and the risks of overreach and hubris offer enduring lessons for today's engineers and professionals as well."

—Robert Westervelt, *Editor-in-Chief, Chemical Week*

"Ralph Landau was one of the giants of the chemical engineering profession. Both experienced and early-stage chemical engineers can appreciate how Ralph created a new business model as well as learn many lessons from his technical and business leadership."

—June C. Wispelwey, *Executive Director, AIChE*

Peter H. Spitz

Primed for Success: The Story of Scientific Design Company

How Chemical Engineers Created the Petrochemical Industry

 Springer

Peter H. Spitz
Scarsdale, NY, USA

ISBN 978-3-030-12316-1   ISBN 978-3-030-12314-7   (eBook)
https://doi.org/10.1007/978-3-030-12314-7

Library of Congress Control Number: 2019930157

This Springer imprint is published by the registered company Springer Nature Switzerland AG
The registered company address is: Gewerbestrasse 11, 6330 Cham, Switzerland

This book is dedicated to a group of largely chemical engineers who worked at Scientific Design Company for periods between its founding in 1946 to the early 1980s, when its management took an immense gamble and effectively lost the company. Those of us who worked there will never forget the Camelot-like atmosphere that prevailed and allowed us to achieve extraordinary results.

Foreword

In 1948, as a recently graduated mechanical engineer, I joined Imperial Chemical Industries, the division of which, situated in North-East England, became known as Heavy Organic Chemicals Division.

ICI, at that time, was wary of the new discipline—chemical engineering—which was beginning to flourish, mainly in the USA. The philosophy in ICI then being to recruit chemists, expecting them to pick up the rudiments of engineering, and engineers, expecting them to learn, on the job, some of the chemistry of the products and processes involved in their particular division of ICI.

A very senior engineer, on a recruiting mission to the nearby university from which I was a graduate, came to tell those of us about to graduate, of the very high level of engineering involved in designing chemical processes, involving designing much of the equipment required, and having them built. All this was new to me, as was ICI, of which I had, until then, never heard. To underline his point, he invited me to visit an ammonia plant nearby.

He and I walked and saw, in detail, the mile-long process, starting with the coke oven plant, to the ammonia reactors, designed to operate at two hundred and fifty atmospheres and requiring very novel engineering at that time. "There you are, young man" said my host at the end of the tour, "it is all high-quality engineering, except what goes through the pipes, and in the reactors."

I was hooked!

What a transformation in the petrochemical industry then ensued! The plastics era took off. Hardly a month went by when there was not an announcement of a new plastic, or fiber, or a new and better process to make them, hitting the news. It was a very exciting time for the chemists, chemical engineers, and engineers of all kinds to be involved.

Scientific Design Company was right in the middle of all this activity. It had found a unique niche for itself. ICI ended up having to license five processes from SD, because, even though ICI had a very large research program on products and processes, SD beat them to the patent office on five of the processes on which they were working!

Chapter 7 of Peter's book sets out very well why SD were more agile than most chemical giants of the time.

ICI in the early 1950s eventually embraced the, then new to the UK, discipline of chemical engineering.

One of the principal reasons, in my opinion, for SD's success and rapid rise to become a major player in the petrochemical world was that the firm recruited, right from the start, very smart chemical engineers. They were an outstanding group of individuals who worked together to form a formidable team. Probably, only DuPont at the time could claim to have such a productive team of chemical engineers.

I would add three other points to Peter's Chap. 7 which applied to ICI. Perhaps some of these points applied to other chemical companies.

- ICI set up a Corporate Laboratory wholly dedicated to "blue skies" research. This, in my opinion, tended the research departments of the ten ICI divisions to take their eye off the ball regarding a search for a better chemical route to the desired product. There grew a tendency to expect from corporate, but not divisional research, a new breakthrough.
- There was little or no effort expended on the engineering process of the existing chemical route. This did not start, in earnest, until the main board of ICI sanctioned funds for the construction of a new plant. There is little time then to develop the engineering process, other than incorporating minor changes, because the urge was to build the new plant.
- Also, there became in the UK, as well as at ICI, a wariness against new processes for chemical routes, because of troubles that arose in the many new nuclear power plants and ammonia plants being built at that time with untried processes.

Peter Spitz, in this book, captures the fascinating story of SD and Halcon/SD, very accurately. The book is a comprehensive and well-written account of those transforming years of the petrochemical world, and the pivotal and unique role that SD played in it.

It was not realistic of Ralph to think that Halcon could become a large operating company. Halcon did not have the necessary staff with the relevant experience to achieve it. Ralph had already achieved one of his goals, that of being renowned throughout the world.

He should have been content with that. He became as well known and respected, worldwide, as his contemporaries at MIT, many of whom had become presidents of major oil and petrochemical companies.

Peter's book is a fitting tribute to the vision and entrepreneurship of Ralph Landau and Harry Rehnberg.

London, UK   Sir Robert Malpas, CBE
                                                                       Member, Main Board ICI (retired)

Preface

In the summer of 1956, I saw an advertisement in the New York Times for a position for a chemical engineer, posted by a firm I had never heard of: Scientific Design Company. At the time, I was a group leader at Esso Engineering, a division of Standard Oil Development Company. I had graduated from MIT with bachelor's and master's degrees and was convinced that chemical engineering would be my chosen profession. At MIT, I had been fortunate enough to have both Professor Warren K. Lewis and Edwin Gilliland as instructors. Lewis, in particular, had made a lasting impression on me.

For the first seven years I designed and started up a number of petroleum refining plants and then felt I wanted to become more involved in petrochemicals, at the time in a high growth mode. I was interviewed by Ralph Landau, one of the firm's founders, and shortly thereafter started to work for the company we would always refer to as SD.

The change from working for one of the largest companies in the world to joining a firm with, at that time, less than one hundred employees was amazing. While I will always feel that my work experience at Esso Engineering was extremely rewarding and enlightening, the atmosphere I became a part of at SD was much more fulfilling: the opportunity to work in an entrepreneurial company that had to live by its wits, while competing with established chemical operating companies and engineering contractors. For a small company with a small laboratory, SD at that time was already engaged in dozens of projects involving a number of chemicals. When a project was secured by Landau or other sales executives, the client did not realize that we often did not have a depth of knowledge of the process upon which the design would be based, but we were resourceful, were able to gain information quickly, and had the chemical engineering skills to come up with a good design. What helped the situation was the fact that many companies wanted to get into petrochemicals production, but could not obtain the needed technology from the established producers. So, they turned to SD, realizing that they had to trust us to help them build a plant that could compete in the new, exciting industry.

For eight years, I worked on process design and process development. At some point, I was put in charge of designing and starting up a phthalic anhydride plant for Witco Chemical without having the slightest idea where to start. I decided to look at reports that American chemical executives brought back from Germany after World War II. That proved invaluable. We used those reports and other material to design the plant and to find an appropriate catalyst. Started up in a cold winter in Chicago, the plant immediately experienced fires and explosions and was hampered by the need to re-weld over a thousand reactor tubes without taking the reactor down. Somehow, we made the plant work and this was my "baptism of fire," in a real sense, as a chemical engineer.

At some point, I was put in charge of process licensing, ending up working directly for Ralph Landau and becoming an officer of the firm. Bob Davis, who had joined SD just before me with a Ph.D. in chemical engineering from MIT, had distinguished himself, developing an ethylene oxide process and catalyst under the direction of Bob Egbert, another of SD's founders. He was put in charge of process development and also became an officer. Davis and I soon recognized the unique nature of SD and believed that we should have "a small piece of the action." When this was refused, we decided to form a consulting firm called Chem Systems.

Over the period 1964 to 1998, Chem Systems became what we believed was the leading petrochemical management consulting firm, with offices in New York, London, Paris, Tokyo, and Bangkok. We also built a laboratory that worked on processes for maleic anhydride from n-butane, propylene oxide, and methanol, the latter using fluidized bed technology. In the 1980s, Chem Systems' consulting arm had a strong strategy and financial practice and was often selected to work in tandem with large Wall Street firms, supporting mergers and acquisitions that were very common during this period, when petrochemical firms had to decide on their long-range goals.

I wrote my first book, *Petrochemicals. The rise of an industry* during this period, motivated to do this because I was interested in interviewing some of the, then already much older, people who had played a role in the events that led to the conversion from coal chemistry to hydrocarbons. With the help of Dieter Ambros, then a high level executive of Henkel, and a good business friend, I spent an afternoon in Mannheim, Germany, with his father Dr. Otto Ambros, earlier an executive of I.G. Farbenindustrie, who headed up a complex that made Buna rubber during World War II at a plant near Auschwitz and was convicted as a war criminal because he had employed "slave labor." Rehabilitated, he told me about his post-war experiences working for W. R. Grace's chemical business, but he also recounted stories of his career at I.G. Farben.

In 2002, I wrote my second book, *The Chemical Industry at the Millennium: Maturity, Restructuring and Globalization*, continuing my involvement and fascination with the global petrochemical industry.

I have periodically been back to MIT, at one time giving a seminar sponsored by Dr. Robert Armstrong, then head of the Chemical Engineering Department, on how to create a consulting firm. When the breakthrough on shale oil and gas came along, I was asked to give a Ted$_x$ speech on how chemical engineers helped to bring about a renaissance in the energy and petrochemical fields after a period when the public had started to think of these as "smokestack" industries.

About two years ago, Alan Hatton, head of MIT's Chemical Engineering Practice School, asked me to speak at the annual meeting attended by the Course X chemical engineering faculty and by graduate students. Alan was familiar with a keynote speech I had given a year earlier in Dubai to the Gulf Coast Petrochemical Association, in which I discussed the historical technology development of the petrochemical industry. As it turned out, both speeches strongly featured the role of Scientific Design Company as a leading source of breakthrough petrochemical research. While I had been away from SD for a number of decades, I could never forget the exciting time that my colleagues and I spent at the firm when the petrochemical industry was in its infancy.

Soon thereafter, it all came together in my mind. The story of Scientific Design and its leader Ralph Landau was compelling, even if it ended badly, with the firm closely avoiding corporate bankruptcy after deciding to become an operating company. I again recognized the important role of MIT in leading the creation of the chemical engineering discipline, which was so crucial in rapidly building a synthetic rubber and aviation gasoline industry that helped the Allies to defeat Nazi Germany and Japan. And I thought about the fact that the petrochemical industry was certainly unique among other heavy industries in disrupting a long-term status quo in raw materials. Over a short time, chemicals that were made from coal, wood, and alcohol would now be made from petroleum and natural gas.

I knew that I could not write this story without knowing more about what happened at Scientific Design after I left to form Chem Systems. Also, I wondered how it was possible for a small, undercapitalized firm with great ambitions to develop breakthrough technologies and thus to outperform the research departments of storied firms like DuPont, ICI, BASF, and Monsanto. I decided to interview as many "alumni" of SD that I could locate. A stroke of luck then allowed me to peruse a number of oral histories of chemical executives who had been interviewed over the years as part of its Oral History Program by the Chemical Heritage Foundation (now called Science History Institute) in Philadelphia.

When writing this book and deciding on what kind of information to include, I opted to provide some details on the actual chemistry describing the petrochemical intermediates and end-products that made up the new industry. At the risk of getting too granular, I decided to provide simplified flow diagrams, illustrating processes. These will be easy for readers with a little chemical background to understand. Perhaps there is some nostalgia involved here, as the chemistry covering petroleum-based molecules reacting to form petrochemical end-products was always a fascinating area for me. I recognize that researchers in the chemical industry now appear to have largely moved on to biochemistry, material sciences, nanotechnology, and other new areas. The heyday of petrochemical process development only lasted for a few decades. It seems worth remembering.

Scarsdale, USA   Peter H. Spitz

The original version of the book was revised: Missed out author corrections have been incorporated. The correction to the book is available at https://doi.org/10.1007/978-3-030-12314-7_15

Acknowledgements

When I started to think about writing this book, I knew that I needed to seek a lot of information about events that took place fifty or more years ago. It was particularly important for me to try and contact people who had worked at Scientific Design Company, recognizing that many of these have already passed away. So, I started to call people I have still been in contact with, interviewed them, and asked them to give me any contact information they might have for other SD "alumni." This turned out to be a productive exercise.

I would like to extend my most sincere appreciation to this very special group. I will start with Harold Huckins, whom I met in 1956 when I first joined SD, and whom I recently interviewed in Chesapeake, Virginia. Huck has maintained an exceptional memory as well as copious files and was certainly my most important source, not only about the early days of SD but also about the eventual downfall of Halcon. Huck held responsible management positions during his career at SD and made numerous important contributions in the development of SD's technologies. His continuing contacts with Bob Egbert, after Egbert left SD, led to Huck's periodically receiving valuable information about SD's early days in the form of Egbert memoirs, which he made available to me. John Schmidt, now living in Princeton, has also been an excellent, indefatigable source, particularly for the period after Oxirane was created. John and Huck tried hard to make the MEG plant work, but to no avail and this eventually led to Halcon/SD's downfall. Martin Sherwin and Marshal Frank provided encouragement and helped me to connect with SD people they had worked with, as well as relating their experiences in SD's laboratory and on sales trips. Larry Nault reviewed some chapters and helped with recollections. Joel Kirman volunteered to draw the diagrams illustrating a number of processes. The following people, in no particular order, who worked at SD, were also very helpful in providing information about their experiences at the firm: Brian Ozero, Bob Hoch, Neil Yeoman, Joseph Porcelli, Ernie Korchak, Joe Porcelli, Alan Sullivan, Barry Evans, Bert Lewin, Alan Peltzman, and Ron Cascone. For Japanese recollections, I relied on James Yoshida, who joined SD as head of the Tokyo office after a number of years at Mitsui Petrochemical, as well as on my friend Ryota

Hamamoto, who was a high-level executive at Sumitomo Chemical. In many cases, I have quoted them in the book. Several of these colleagues saved mementos of their time at SD and made them available to me. I am particularly grateful for the use of sales brochures printed by Scientific Design Company in 1962 and 1968.

Sir Robert (Bob) Malpas, ex chairman of Imperial Chemical Industries and, for several years afterward, president of Halcon, provided and reviewed information on SD/Halcon during the period when he was an executive there.

Jon Rehnberg, son of one of the three founders of SD, was contacted by Marshall Frank, with whom he had worked, met with us near his home in Connecticut to provide information on SD's early days, as recalled by his father, as well as on his time as president of SD, before its sale to Texas Eastern Corporation.

Other people who contributed important information, in their case mostly about the fateful MEG project, include Jim Terangelo (who also provided the names of other people of interest), Morris Gelb, Ben Binninger, Wayne Kuhn, Wayne Wentzheimer, and Ed Zenzola, all of these coming from the Arco Chemical/Oxirane side.

I also received considerable help on petrochemical lore from my old friend Joe Pilaro, who spent much of his career at National Distillers and Chemicals Corporation and was a part of the growing petrochemical industry.

Tom Lewis, an ex Monsanto executive, who served on the Executive Committee of the Chemical Heritage Foundation with me, must be mentioned here. It was a tremendous chore getting releases signed for some photographs in the book that were taken in the 1950–1970 period by chemical companies no longer in existence today. When I became desperate regarding a photograph of a maleic anhydride plant built by Monsanto in Newport, England, my email found Tom on a family trip to Cambodia. He immediately set to work contacting one of his colleagues (Tom Potter), who had gone with Solutia when this firm was spun off from Monsanto decades ago. Unfortunately, this did not work out, but I was able to get a release from Pfizer company covering a picture of a fumaric acid plant that SD had designed for that firm. Similar exercises, getting releases from spun-off, merged, or now defunct companies, were among the most difficult tasks in getting this book ready for publishing.

Jeffrey Plotkin, at one-time head of Chem Systems' Process Evaluation and Research Planning (PERP) program, has been of great help in getting the chemistry of the various processes right.

David Hunter, ex editor of *Chemical Week* magazine and now with McKinsey and Company, reviewed some chapters and was instrumental in my decision to add flow diagrams to the text.

Of great value to me were the oral biographies of a number of petrochemical executives available from the Science History Institute (formerly Chemical Heritage Foundation) in Philadelphia. Of particular importance was the oral biography of Hal Sorgenti, ex CEO of Arco Chemical, which provided a great deal of information on how Halcon/SD's last project went awry, leading to the buyout of Oxirane. Great thanks to Ann Sorgenti, who facilitated my contacts with Hal, who unfortunately passed away during the time I was writing the book. Other important oral biographies included those of Gordon Cain (Conoco Chemical, Vista Chemical), James Roth (Monsanto), Jim Fair (Monsanto), Arthur Mendolia (Oxirane), Jim Idol (Sohio), Delbert Meyer (Amoco Chemical), and William Asbury (Exxon). Elsa Atson and her staff at the Science History library were very helpful in sending requested reference material.

My sincere thanks to Hilda for being very supportive over the months that it took to collect and catalog all the information I needed, to do without a computer some of the time and to recognize that a lot of patience is required when a home office becomes the site of a creative endeavor.

Contents

1   Introduction   1
    References   4

2   The Global Chemical Industry Is Poised for Change   5
    2.1   Chemical Companies Pursued Different Directions   10
        2.1.1   DuPont   10
        2.1.2   Monsanto   11
        2.1.3   Celanese   11
        2.1.4   Allied Chemical   11
        2.1.5   Union Carbide (UCC)   11
        2.1.6   American Cyanamid   12
        2.1.7   Dow Chemical   12
        2.1.8   Rohm and Haas   12
        2.1.9   Standard Oil Company (N.J.) (Exxon)   12
        2.1.10   Shell Chemical   13
        2.1.11   Ethyl Corporation   13
        2.1.12   In Europe: Little Interest in Petrochemicals   14
        2.1.13   Chemical Engineering: The Enabler of an Industry   15
    References   23

3   Fuels and Chemicals Research Helps Win World War II   25
    3.1   Companies Start Looking at Hydrocarbon Feedstocks   28
        3.1.1   Union Carbide   28
        3.1.2   Shell Chemical   29
        3.1.3   Standard Oil (N.J.) (Exxon)   30
        3.1.4   Dow Chemical   32
        3.1.5   Other Companies   32
    3.2   Petrochemical-Based Polymers Started Their Phenomenal
            Growth During World War II   33
    3.3   German Chemistry Was Essential for U.S. War Effort   34

     3.4   A Massive Construction Program Helps Win the War   37
     References   41

4   **Three Entrepreneurs Join Forces**   43
     References   55

5   **Circumstances Are Perfect: A New Industry Grows Rapidly**   57
     5.1   Why So Many Firms Became Petrochemical Producers   61
     5.2   Europe Goes Petrochemical—A Little Later   65
     5.3   Great Britain   66
     5.4   Germany   67
     5.5   France and Italy   69
     5.6   Japan   70
     References   72

6   **Scientific Design Company Becomes Successful**   73
     6.1   Scientific Design Becomes a Developer of Petrochemical
            Technology   75
     References   94

7   **The First Major Invention: The Mid Century Process**   95
     References 106

8   **More Inventions: Nylon Intermediates, Isoprene, Propylene**
   Oxide, Methyl Methacrylate 109
     8.1   The Next Breakthrough: High Yield Cyclohexane
            Oxidation 112
     8.2   Isoprene 118
     8.3   Epoxidation of Propylene 120
     8.4   Methyl Methacrylate 123
     References 124

9   **Other Firms: A Period of Breakthrough Inventions** 125
     9.1   UOP Platforming 127
     9.2   Monsanto's Acetic Acid Process 129
     9.3   Sohio Acrylonitrile Process 131
     9.4   Ethylene Oligomerization and Shell's SHOP Proess 132
     9.5   National Distillers: Vinyl Acetate (VAM) 135
     9.6   Oxychlorination of Ethylene 136
     References 137

10 **The Leading Research and Engineering Firm**
    in Petrochemicals 139
     10.1   The Diverse Nature of the Projects and the Exciting
            Atmosphere that Prevailed 142
     10.2   Third Party Processes Available Through Scientific Design 145

10.3 Scientific Design Diversified into Coal-Related Work
as the Oil Shocks Impact the US Energy Situation 146
10.4 Scientific Design "Wrote the Book". 147
10.5 Linking Up with MIT 148
10.6 The Stage Was Set 149
References 149

11 Oxirane and the Creation of Halcon International 151
11.1 The Creation of Oxirane 154
11.2 Construction of the Bayport Plant 155
References 164

12 A Reach Too Far 165
12.1 Background of the MEG Project 166
12.2 Plant Startup and Operation: Serious Problems 172
12.3 The Oxirane Buyout 177
References 180

13 An Ending. 181
13.1 Acetic Anhydride: The Last Success 182
13.2 The Polyvinyl Alcohol (PVA) Problem 184
13.3 Halcon/SD and Oxiteno Look at Brazilian Alcohol
as a Feedstock 185
13.4 Government Work Helped, but not Enough 185

14 Epilogue 187
14.1 Ralph Landau 187
14.2 Propylene Oxide: The Legacy of a Significant
Achievement 189
14.3 Scientific Design Company: Still Competing on Ethylene
Oxide. 190
14.4 The Petrochemical Industry 191
14.5 Chemical Engineers Turn to Other Challenges 193
14.6 The Unusual History of Petrochemicals 195

**Correction to: Primed for Success: The Story of Scientific Design
Company** C1

About the Author

Peter H. Spitz has written and spoken extensively about the petrochemical industry. He developed a strong interest in chemistry, receiving bachelor's and master's degrees in chemical engineering from MIT. He worked at Exxon (formerly Std. Oil Company (N.J.)) and at Scientific Design Company before forming Chem Systems Inc., an international management consulting firm. He holds nine patents covering processes in the chemical and fuels industry. He is on the Executive Committee of the Science History Institute.

Chapter 1
Introduction

Abstract Less than a hundred years ago, the U.S. chemical industry was poised to undergo dramatic change, though there was little awareness that a hitherto stable industry was about to be disrupted. But momentous changes took place as World War II introduced a number of new technologies and the industry became a major user of petroleum-based feedstocks that could be converted to organic chemicals and plastics. The circumstances that occurred and the companies that played a role in this transformation are the subject of this book. While operating companies were the major actors, independent research and engineering companies also played an important role. And one firm, Scientific Design Company, headed largely by chemical engineers from MIT, became highly successful by developing a number of technologies that helped shape what became known as the petrochemical industry.

Let's consider the United States in 1936, when Franklin Roosevelt was re-elected to a second term. The country had experienced the great economic Depression, millions of people were out of work, and the government was spending money to create jobs and alleviate economic distress. A boom period after World War I had allowed many families to build new houses, purchase automobiles and live a good life, but in 1929 everything had started to go bad and now the economic situation was grim, with people having little discretionary income, dramatically reducing the demand for durable consumer goods. People were hanging on, hoping for better times.

Fortunately, there was ample fuel for cars and trucks, with oil production rising steadily and refineries converting the crude oil into gasoline, diesel fuel and industrial heating oil. Gasoline was cheap: six gallons to the dollar! In Texas, Oklahoma and some other states, natural gas was also plentiful and inexpensive, used as an industrial fuel and for home heating and cooking. The bulk of U.S. consumers were using Town Gas, a low BTU fuel gas produced from coal and distributed in pipelines to industry and households. Natural gas was not well known in other parts of the country and certainly not thought of as having any use for chemical manufacturing.

As drivers challenged car manufacturers to build more powerful car engines, the refining industry developed techniques that could produce a gasoline to match the performance of these engines, i.e. to make what became known as "high octane" gasoline. New oil "cracking" technologies not only created more powerful gasolines,

© Springer Nature Switzerland AG 2019
P. H. Spitz, *Primed for Success: The Story of Scientific Design Company*,
https://doi.org/10.1007/978-3-030-12314-7_1

but also made volatile by-products that looked attractive to researchers as feedstocks
for the production of other high octane gasoline components as well as chemicals.
Chemists also found that the naphthas recovered from fractionation of certain crude
oils, as well as from cracking operations, contained substantial amounts of aromatic
hydrocarbons which not only had a high octane value, but could also be converted to
organic chemicals and polymers. The result was that in the search for more powerful
gasolines, scientists found new areas of research that would herald the advent of the
petrochemical era.

Over the years, chemical companies had built large plants to make organic chemicals from coke oven byproducts and from fermentation alcohol, as well as man-made
fibers based on cellulose from wood (i.e. rayon, acetates). In 1913, Fritz Haber and
Carl Bosch, in Germany, had succeeded in synthesizing ammonia, a key material for
fertilizers, explosives and chemicals, using high pressure equipment and a catalyst, a
development for which they earned a Nobel Prize. Then, in the 1920s, the company
later known as Union Carbide discovered how to produce ethylene from the natural
gas found near its home base in West Virginia. The company built a small plant
to produce ethylene glycol for its antifreeze business and produced other ethylene
derivatives. But it was ahead of its time.

In 1931, Dr. Leo Baekeland had successfully manufactured a family of soft and
hard plastics by reacting phenol or urea with formaldehyde. In 1936, telephone
headsets and dinner dishes were made from this type of plastic. Also, acrylic plastics, which were generally clear and transparent, could be produced from alcohol-
based acetone. In the U.S. and Germany, chemists developed technology to produce
polyvinyl chloride (PVC), a new rigid plastic material, from coal-based acetylene and
chlorine for use in rigid piping. Early work in both countries also led to the discovery
of polystyrene, another clear material with considerable promise. But these so-called
thermoplastics were produced in small quantities and were almost unknown to the
public. And there was no polyethylene.

In the 1930s, research work at DuPont under Dr. Wallace Carothers had successfully invented a synthetic fiber that became known as nylon. German research
resulted in a polymer very similar to Carothers' nylon, while British scientists carried out parallel work to Carothers' experiments, synthesizing a polyester monomer
suitable for the production of another type of synthetic fiber.

Automobile and truck tires in the 1930s were produced exclusively from rubber,
imported largely from Malaysia. The demand for this raw material kept rising rapidly,
causing researchers in the 1930s in the U.S. and Germany to explore how to make
a synthetic rubber that would have properties similar to the imported material. But
little success was, at first, achieved.

In sum, there was a growing amount of chemicals research based on hydrocarbons
carried out in the 1930s, but little commercialization of this research by the time the
U.S. was attacked at Pearl Harbor.

It was a different story in Germany, which was preparing for the next war and,
by 1936, was already well along in building plants to make tank, automobile and
aviation fuels, as well as synthetic rubber. These key war materials were all based
on coal or lignite, which were the only carbon-based fuel indigenous to Germany

and much of Europe. The German chemical industry, a group of companies known as I.G. Farben industrie, was the leader in the area of chemical synthesis. Hitler commanded these firms to develop and commercialize technologies that would make Germany independent from imports of crude oil and rubber, and in this Germany soon succeeded.

Germany and the United States had different engineering disciplines when developing chemical technology for industrial applications. Chemical Engineering was not then taught in German universities. Instead, chemists worked with mechanical and other engineers when designing and building chemical plants, such as ammonia factories. In the U.S., MIT and other colleges like Cornell, Princeton, Minnesota and Michigan developed strong chemical engineering departments that offered courses in what they called *unit operations*, such as Distillation, Absorption, Heat exchange, Crystallization, etc. When the U.S., post 1941, had to rapidly build new plants based on new chemical research mostly still in the pilot plant stage, chemical engineers were ready to put their knowledge to work.

When the war was over, many of the plants that had been built during the conflict would no longer be needed to make war materials. But some of the products they manufactured could now find uses for both durable and disposable consumer products: housing, cars, appliances, etc. These products were mostly still unknown to the population, but their usefulness was apparent. And a growing demand for these products, together with a booming post-war economy, spurred the growth of the new products known as petrochemicals: PVC for piping, shower curtains and automobile seats: polyethylene for packaging and toys: polystyrene for drinking glasses and foam insulation, nylon for women's stockings, carpeting and hard plastic parts, polyester for garments, synthetic rubber for tires, etc. The demand for these products became so strong that the traditional chemical companies could not keep up with demand.

The global chemical industry before World War II was, in many ways, a very clubby business. Many of the large chemical firms in Europe and to some extent in the U.S. had for a long time participated in cartels for a number of products (e.g. soda ash, ammonia, explosives), setting prices, and limiting competition between these firms. A cartel also agreed among its members regarding construction of new plants so as to avoid excessive competition due to overcapacity. The U.S. government eventually took steps to prohibit U.S. firms from joining cartels, but tacit cooperation continued for some time. The strategy of limiting the construction of new plants no doubt contributed to the decision of traditional producers like DuPont, Union Carbide, Allied Chemical and Monsanto to go slow in building or expanding petrochemical plants after the war, in spite of rapidly rising product demand. Similarly, they decided not to license competitors with their closely held technologies. This opened the door for other companies to enter the petrochemical industry if they could find sources for process knowhow.

This, in fact, was the opportunity identified by three chemical entrepreneurs, Ralph Landau, Harry Rehnberg and Bob Egbert, who founded Scientific Design Company with the goal of developing technologies that new industry entrants could use to enter the burgeoning petrochemical industry. It was an opportunity based on the changes and disrupting growth occurring in the industry and the need for new industry entrants to obtain technologies that the entrenched existing producers would not license or did not have.

So, in many ways, the groundwork was set for the introduction of the petrochemical era in the United States and thereafter in Europe and Japan. This book will describe the role that Scientific Design played in this exciting creation of an industry, developing and commercializing five important, patented chemical processes over a period of twenty years and identifying and obtaining sublicenses for many other technologies they could offer to new entrants. During this period, SD designed and, in many cases, built 275 petrochemical plants in 30 countries.[1]

The transformation of the organic chemical industry, once largely based on coal, to an entirely different and much larger one based on petroleum and natural gas was arguably unique in the history of industrial manufacturing. How this happened, and how the founders of Scientific Design Company showed that there will always be a role for entrepreneurs to become successful in an arena populated by large companies will be one theme of this story. Another is the transformative application of chemical engineering for inventing a number of new technologies, with MIT receiving a great deal of the credit for leading this discipline. This book is about the rapid creation of an entire industry and how it grew at a breathtaking pace to become, for a while, the most desirable place to be. A quotation frames the thesis of this book:

> In a very real sense, Landau, the petrochemical industry and the chemical engineering profession grew up together and nurtured each other….the petroleum-based chemical industry and the chemical engineering profession were in their infancy…..the most attractive alternative to coal was oil, which the country had in abundance…the chemical companies therefore began to develop both continuous and automatic processes for refining petroleum and gasoline and also the processes that would eventually produce petroleum-based raw and intermediate materials for other uses.[2]

References

Ralph Landau (1978) *Halcon International, Inc. An Entrepreneurial Company* The Newcomen Society in North America, New York, 16
Ralph Landau (1994) *Uncaging Animal Spirits* the MIT Press, Cambridge, Mass

---

[1]Landau (1978).

[2]Landau (1994).

Chapter 2
The Global Chemical Industry Is Poised for Change

Abstract The state of the pre-World War II chemical industry is described. Organic chemicals and plastics were based on coke oven chemicals, ethyl alcohol and wood cellulose. It was an industry often connected to international cartels, headed up by I.G. Farbenindustrie in Germany. The main activities of companies such as DuPont, Monsanto, Union Carbide, Dow and Celanese are briefly covered. The advances being made in chemical engineering during the prewar period, and the cooperation between MIT and other universities with DuPont, Standard Oil (N.J.) and other firms are discussed.

The U.S. Chemical Industry in the 1930s was one of the largest manufacturing industries in the country, ranking with steel, glass, automobiles, aluminum and cement as a key part of the country's manufacturing base. It was dominated by a several large firms, some of which had undergone consolidations through acquisitions. Many smaller chemical companies were still run by their founder or his successors. Large companies had research laboratories that came up with patented or unpatented innovations and they purchased technologies to branch out into new, hopefully promising areas. With a few exceptions in the inorganic chemical area, chemical plants were relatively small in size, compared to many plants today and they generally used what is called "batch processing". In this mode, chemicals are produced in small reactors in a manufacturing cycle that involves filling a reactor with ingredients, adding a catalyst, heating to the desired temperature where the reaction occurs, cooling and separating the end product from byproducts and waste via distillation or other means, cleaning out the reactor and starting over again. There was also some "continuous processing", where relatively larger plants operated around the clock making the desired product, shutting down only when cleaning was required or when there was a malfunction. Soda ash, chlor-alkali, and ammonia were, since the Nineteenth Century, produced in this manner by large U.S. firms like General Chemical Company, Michigan Alkali Company and Solvay Process Company, making inorganic chemicals like soda ash. DuPont used continuous processing in its explosives business.

Organic chemicals were largely the province of German firms, who were active in researching catalytic processes, in contrast to U.S. companies, which did less chemi-

© Springer Nature Switzerland AG 2019
P. H. Spitz, *Primed for Success: The Story of Scientific Design Company*,
https://doi.org/10.1007/978-3-030-12314-7_2

cal research than the Germans. The decision by DuPont management to fund research in the late 1927s in three important organic chemical areas could be considered a milestone.

There are obvious similarities between oil refineries and chemical plants, when continuous processing is involved. In the 1930s, the petroleum refining industry was well along with continuous processing, as oil was distilled into various fractions, including light hydrocarbons, gasoline, kerosene, diesel fuel, heating oil, asphalt, etc. Some of these streams were further processed and treated in reaction vessels, for examples to remove sulfur-based impurities. Refinery processing equipment, such as distillation columns, absorption towers, reactors and heat exchangers, was later used in the manufacture of petrochemicals. Thus, oil refining technology, including attendant equipment design techniques, was useful and relied upon when chemical engineers started to design petrochemical plants.

Chemicals are generally classified as either organic or inorganic, the former being chemicals that are based on carbon. In the 1930s, the original sources of carbon were almost exclusively coal, alcohol, or wood. The manufacture of iron and steel had, for centuries, depended on the production of metallurgical coke from coal in so-called coke ovens, where the coal was heated to high temperatures under reducing conditions, which produced coke suitable for steelmaking, but also produced large amounts of gases, liquids and tars containing a variety of chemicals (e.g. ammonia, benzene, naphthalene, phenol and reactive hydrocarbons like ethylene). These chemicals were, in part, condensed, separated and sold to chemical companies for further processing. Coal was also the raw material fed to electric furnaces to produce acetylene, a highly reactive molecule that could be converted to such chemicals as vinyl chloride, vinyl acetate, methyl methacrylate, and others. Ethyl alcohol, obtained via fermentation, was used as a raw material to produce ethylene, a number of ethylene derivatives and higher ethylene-based chemicals (e.g. oxo alcohols). Ethylene dichloride and dibromide for tetraethyl lead (TEL) were originally made from ethylene derived from ethyl alcohol. In Brazil, before World War II, Coperba built an entire chemical complex bases on alcohol from sugar cane that produced ethylene as a raw material for a number of organic chemicals (alcohols, ketones, etc.).

The cellulose contained in wood was used to make so-called "man-made fibers", including rayon and acetate fibers. Tennessee Eastman Corporation manufactured methanol, methyl acetone, and various byproducts as well as acetic acid through the dry distillation of wood. It started production of acetic anhydride and became a large manufacturer of cellulose acetate yarn and cellulosic plastics in the early 1930s. Celanese similarly became a producer of acetate fibers and plastics.

– It is therefore significant to realize that less than a hundred years ago, the organic chemical industry was almost completely based on coal, alcohol and wood. Considering today's typical 5000 ton per day methanol plants based on natural gas, makes it difficult to recognize that this important chemical now used for the production of huge quantities of formaldehyde and acetic acid was then entirely obtained from trees. Similarly, today's typical large polystyrene plant based on petroleum-derived benzene would probably have needed the production of byproduct benzene from all of the coke ovens at the height of domestic steel production. Fortunately for the chemical industry, hydrocarbons soon provided an almost unlimited source of benzene, while different methods to produce iron and steel resulted in the eventual shutdown of coke ovens, to the benefit of nearby residents who suffered from the pollution inherent in this operation.

It is instructive to ask why companies did not look to petroleum fractions to make the chemicals they were then producing from coal or alcohol. When oil refineries started to use thermal cracking processes to increase gasoline yields, they also produced as byproducts appreciable quantities of ethylene, propylene and other reactive hydrocarbons well suited for organic chemicals production. But oil companies in the 1920s and early 1930s had little interest in making chemicals. For quite a while, the cracker off-gases were not of interest for further processing and were burned as fuel. However, three advances in refinery processing in the 1930s resulted in a major change in refinery operations, largely caused by the rapidly rising demand for high octane naphthas. These advances also significantly increased the amount of olefins and aromatics produced, leading to a recognition that these could become the basis for a different kind of chemical industry. The most important invention resulted in the fixed bed catalytic cracker based on an invention by Eugene Houdry, a French scientist. In the reactor, heavy oil fractions were catalytically "cracked" to produce lighter cuts, predominantly in the gasoline boiling range, with light reactive hydrocarbons, including ethylene, propylene and butylenes as byproducts. Use of a silica alumina catalyst promoted a reaction that produced a considerable amount of molecular branching, resulting in gasolines with higher octane number units.[1] During the cracking operation, coke was deposited in the pores of the catalyst, making it progressively more inactive. The reactor was therefore periodically shut down and the catalyst "regenerated" by burning off the coke. The reactor was then placed back in operation. The process was piloted by Socony Vacuum (later Mobil Oil Company) and commercialized in the form of a 15,000 barrels per day cracker by the Sun Company (now Sunoco Inc.) in 1937.[2] Other companies followed, and a large scale plant to produce the synthetic silica alumina catalyst used in the process started up in 1940. Unique at the time, the Houdry fixed bed reactor was operated in a cyclical manner, with a fixed period of reaction for gasoline production followed by a period of catalyst regeneration via carbon burn-off.

Shortly before the start of the war, Dr. Warren K. Lewis and others at MIT, in work sponsored by Exxon[3] developed an improved system for converting heavy oils to gasoline that was termed fluid catalytic cracking. This operation used a reactor and a regenerator, combined with piping that allowed spent catalyst from the reactor to be continuously withdrawn and sent to the regenerator where the coke on the catalyst

---

[1] Pigford (1976).

[2] Spitz (1988).

[3] In this book, the firm that was originally called Standard Oil Company (N.J.) will be referred to under the name *Exxon*.

was burned off and the regenerated catalyst sent back to the reactor. Both reactor and regenerator maintained a "fluidized" bed that emulated the way a liquid would work. Eventually, all catalytic crackers were built using this mode.

The second development was the conversion of low octane (high straight chain aliphatic) naphtha, as distilled from crude oil, into high octane gasoline. Research by Exxon, Chevron, and others resulted in a fixed bed catalytic process called *Hydroforming*, which produced a gasoline-fraction naphtha with a high content of branched molecules as well as benzene, toluene and xylenes (BTX aromatics) with high octane rating. The technology used in this process largely came from Germany in an exchange program discussed later in this book.

The third development was a process to extract the aromatics from their presence in the mixture of straight chain and branched aliphatics and separating the three aromatics components to produce large quantities of benzene, toluene and xylene. This was done with a sulfur dioxide extraction system invented by a Romanian chemist Lazar Edeleanu, who is sometimes credited with inventing the modern method of refining crude oil.

The three processes described above are not only important by themselves, but also because they employed production techniques later used in many petrochemical processes.

Refinery operations therefore were starting to produce large amounts of "feedstocks" for organic chemicals production. Moreover, natural gas already produced in large amounts on the U.S. Gulf Coast for heating and power generation purposes, contained large amounts of ethane and other low boiling aliphatic hydrocarbons that are very suitable to be converted to the same reactive types of feedstocks. Researchers were therefore also working on high temperature thermal cracking of these hydrocarbons into reactive olefins. This operation was the harbinger of today's immense ethylene plants.

In addition to research performed in oil companies' laboratories, there was also research carried out by an engineering company originally set up by a group of these firms to do independent research in petroleum related areas: Universal Oil Products, later known as UOP. This company was set up to license its technology to any company interested in acquiring it, thus earning funds to carry out more research. This is important to this story in that UOP established the fact that engineering firms could develop technologies that they would license to operating companies. In the case of UOP, which paid for all the research, it would seek licensing income from as many firms as possible. This was an acceptable situation for licensees since none of them could use the technology to gain an advantage over competitors, given the very large number of companies using or having access to use the technology. That this was a different situation for Scientific Design Company as we will see later.

In the 1930s, researchers recognized that petroleum and natural gas could provide feedstocks for organic chemicals production, but they tended to look with greater interest to propylene, butylenes, benzene and toluene rather than to ethylene. What kept companies from considering ethylene as a key raw material was the fact that there was no large ethylene demand, with polyethylene yet to be commercialized. Little more than a decade later, polyethylene became the largest selling polymer in the world. Yet is was barely known by the start of World War II, when the first small plant for polyethylene production was built to produce material for radar cables and other highly specialized uses.

In Germany, Adolph Hitler was preparing for war. He recognized that its army, navy and air force would need huge amounts of different types of fuels as well as rubber for tires and tank treads and that neither crude oil or rubber would be available as raw materials as both were imported from areas far away. All Germany had was coal and lignite and these would need to be used for fuels, chemicals and synthetic rubber production. And so it turned to I.G. Farben and its heralded chemical industry and directed it to develop coal-based technology that would make Germany independent of crude oil and rubber imports.

In the early 1930s, Germany's chemical industry, which was the most advanced in chemical technology, was developing two key processes that were able to meet the challenge. In the Fischer-Tropsch process, coal-based synthesis gas (a mixture of carbon monoxide and hydrogen) was passed over a catalyst under high pressure and temperature conditions to produce a mixture of hydrocarbons in the gasoline and diesel fuel range as well as a number of oxygenated organic chemicals also suited for the war effort. The other area of research was the production of different types of synthetic rubber from styrene, butadiene and acrylonitrile (BunaS and BunaN). There had been a technology exchange between Exxon and some I.G. Farben (*Interessengemeinschaft Farbenindustrie Aktiengesellschaft*) firms in the 1930s that gave each side some very useful information. It allowed U.S. companies to learn how to make synthetic rubber for tires while providing the German firms with useful information on a specialized synthetic rubber based on butylene and butadiene suitable for tire inner tubes that the U.S., firm had commercialized. Other German technologies, such as acetylene production and hydrotreating were also disclosed to Exxon under the exchange arrangement, while Germany received the technology to make tetraethyl lead (TEL0, though this was not part of the I.G. Farben/Exxon exchange.

It is hard to appreciate today the extent to which German firms controlled the world's organic chemistry before World War II. The I.G. Farben cartel had been created in 1925 to rebuild Germany's chemical industry and to prevent duplicate production among German producers, to set pricing and thereby to increase mutual profits and keep out competitors. It included BASF, Bayer, Hoechst and several other firms and was headquartered in Frankfurt. Then, the cartel was expanded to include a number of large foreign firms. Eventually, it held a sometimes controlling interest in 379 German and 400 foreign firms, including what became General Aniline & Film Corporation in the U.S. Other European companies in the cartel included I.C.I. in England, Pechiney, Kuhlmann and Rhone Poulenc in France, Nobel in Scandinavia and Montecatini in Italy.

As discussed in more detail below, several U.S. firms, including DuPont, Allied Chemical and others also cooperated with IGFarben, creating a cartel-friendly atmosphere in which companies operated without substantial competition, in a number of chemicals.

2.1   Chemical Companies Pursued Different Directions

While there was no petrochemical industry in the 1930s, the U.S. chemical industry was vibrant, more active in inorganic chemicals than in organic. Companies had different areas of focus and R&D and conducted thriving businesses with then existing technologies. Research was largely focussed in areas of the companies' chemical business, i.e. there was little so-called "blue sky" research. "Petrochemicals" was not a known or described business area, except at Union Carbide, which became the first U.S. company to produce ethylene from a hydrocarbon.

The activities of a number of important U.S. chemical firms active at that time and before there was much interest in hydrocarbon-based chemicals is briefly described below.

2.1.1   DuPont

Established in 1802 as a manufacturer of gunpowder, DuPont became the leading chemical company in this country, broadening into nitrocellulose, nitroglycerine, other cellulose chemistry, including Rayon, lacquers and many other products. It invested in the early automobile industry, buying stock in General Motors. Its Central Research Laboratory became the source of many chemical inventions. Hiring Wallace Carothers in 1928 to work on synthetic polymers, DuPont commercialized Neoprene, as well as the synthetic fiber *Nylon* just before the outbreak of World War II. From acetylene, it made methyl methacrylate and converted it to a transparent plastic it called *Lucite*, a product similar to Rohm & Haas' *Plexiglas*. Dupont's joint venture with Imperial Chemical Industries created a worldwide semi-monopoly in synthetic paints.

In many ways, DuPont was the leading U.S. chemical firm of its time. DuPont engineers "are known to posterity for having developed nylon in the 1930s and then plutonium, crucial to the Manhattan Project. They changed the culture of their company, drawing it into the public sphere, while responding to a massive demand for consumer goods".[4] It worked on polyethylene in the high pressure Ammonia Department and started production of this polymer during the war.

In contrast to other firms, DuPont, the preeminent U.S. chemical firm, had a large research department, allowing its scientists to carry out some "bootleg research" not tied to the company's product lines, and rewarding them (e.g. Carothers) with sometimes substantial bonuses.[5] In general, DuPont's breakthrough inventions were more in the area of polymers than in organic intermediates.

---

[4]Pap A. Ndiaya. *Nylon and Bombs. Dupont and the March of Modern America* The Johns Hopkins University Press. Baltimore, Md. 3.

[5]Hounshell and John Kenly Smith. *Science and Corporate Strategy* Cambridge University Press. New York, NY. 305.

2.1.2 Monsanto

Founded in 1931 by John Queeny, who came from the pharmaceutical industry, Monsanto first produced artificial sweeteners, vanillin and aspirin. It later started producing sulphuric acid, phosphorus and a number of organic chemicals, agrochemicals, and rubber processing chemicals. The company produced polyvinyl butyral in the mid-1930s and built a styrene and polystyrene plant in Texas City, Texas just before World War II broke out.

2.1.3 Celanese

Cellulose acetate was first synthesized in 1913 in Switzerland and was used to produce films and yarns. Founded as the American Cellulose and Chemical Manufacturing Company, Celanese produced cellulose acetate from wood pulp for the manufacture of fabrics and yarns as alternatives to silk. This material is still being used to make cigarette filters.

2.1.4 Allied Chemical

In 1920, five chemical companies including Semet-Solvay, General Aniline and Dye and General Chemical were merged to form Allied Chemical and Dye Corporation, which for a while was the largest U.S. chemical company as a supplier of basic chemicals, including soda ash, indigo dyes and other colorants, phthalic anhydride and other organic chemicals.

2.1.5 Union Carbide (UCC)

Founded in the late 1800s as a manufacturer of calcium carbide and acetylene, the company began manufacture of what later became ethylene derivatives from acetylene. It started research in the 1930s on making ethylene from natural gas liquids, building a small ethylene and ethylene glycol plant to make antifreeze liquid, which had originally been made from acetylene. In Clendening, W. Va., near natural gas deposits, UCC built the first small domestic ethylene plant, using ethane feedstock. This was also the site for making other organic chemicals, including vinyl chloride. Both UCC and B. F. Goodrich discovered how to produce and process polyvinyl resins (PVC) with so-called plasticizers, which softened the resin, allowing pre-war production of such items as shower curtains and films.

2.1.6   American Cyanamid

Cyanamid chemistry using coal-based acetylene, was used to produce many nitrogen-based chemicals, such as acrylonitrile, melamine (used for plastics) and acrylamide as well as a number of pharmaceuticals and chemical additives. The company also produced nitrogen-based fertilizers and phosphates in Florida. After World War II, it became a manufacturer of acrylic fibers.

2.1.7   Dow Chemical

Herbert Dow, who invented a method of extracting bromine from Michigan brine, started a company in 1902 to produce bleach and potassium bromide. Dow Chemical soon diversified its product line to make chlorine, caustic, phenol-based dyestuffs and magnesium metal. Dow produced both ethyl chloride and ethyl bromide from alcohol-based ethylene for making tetraethyl lead for octane improvement.[6] It shifted production of bromine and magnesium to Freeport, Texas, where in 1941, it also started production of ethylene from natural gas liquids as well as ethylbenzene, styrene, polystyrene and vinyl chloride. Dow had entered the manufacture of styrene and polystyrene in the late 1930s.

2.1.8   Rohm and Haas

Founded by a German chemical executive, Otto Haas, the firm started manufacturing methyl methacrylate (MMA), an important paint ingredient, from acetylene in the 1920s. It later also produced acrylic acid and acrylate resins and methyl amines. MMA is the raw material for *Plexiglas*, a transparent plastic used as a substitute for glass.

2.1.9   Standard Oil Company (N.J.) (Exxon)

Recognizing the potential of propylene as a feedstock (raw material) for chemicals production, Std. of N.J. built a plant in the 1920s at its Byway, N.J. refinery to make isopropyl alcohol, using sulfuric acid that reacted with propylene, followed by hydrolysis. This plant made 20,000 lbs. per day of isopropyl alcohol, some of which was converted to 3000 lbs of acetone. It also entered a joint venture with Ethyl Corporation to make ethyl chloride for the manufacture of tetraethyl lead (TEL) for gasoline. Also, before the war, it developed technology to catalytically

---

[6]Graham (1939).

reform petroleum naphtha into aromatics and branched chain aliphatics, a process termed *Hydroforming*. Hydrogenation technology was actually obtained from I.G. Farbenindustrie in a technology exchange agreement that was part of cooperation between the two entities started in 1927. This is further described in Chap. 3.

2.1.10 Shell Chemical

Recognizing the potential of hydrocarbon feedstocks, Shell Chemical's laboratory in Emeryville, California developed in the 1930s technology to produce butyl alcohol and methyl ethyl ketone from propylene. It also pioneered a process to make glycerin via allyl chloride, but did not build a glycerin plant until after the war. Another product made by Shell before the war was ammonia for California's agricultural industry. The firm imported two small European plants to start ammonia production in the early 1930s.

2.1.11 Ethyl Corporation

This firm was founded in 1923 as a 50–50 joint venture between General Motors and Standard Oil (NJ) to manufacture tetraethyl lead (TEL). Since neither companies had the background to run a chemical operation, the venture hired DuPont to design and operate the first plant, which started up in the late 1920s.

All of these companies and a number of others, mostly smaller, operated in the 1930s and generally resumed normal operations at the conclusion of the conflict. As noted earlier, parts of the global chemical industry were used to operate in a form of oligopolistic competition, though the U.S. government periodically took actions to break up cartels or to sue companies for price fixing. The following excerpt is taken from an earlier book.[7]

> The manufacture of heavy organic chemicals, fertilizers, dyestuffs, and a number of other chemicals during the period between the wars was a closely held business, with each segment controlled by a limited number of companies that wanted to keep things the way they were. Cooperation between internationally oriented companies producing similar lines of chemicals was extensive and the business was conducted in an atmosphere of "noblesse oblige" between the large companies that controlled specific product areas. (e.g. Soda Ash, Dynamite) Market sharing on a local and international basis was normal practice in these cases, so that competition in the true sense of the word was often non-existent in most of the countries involved in cartel arrangements.

This was the industrial and pharmaceutical chemical world as it existed in those prewar years. It was a world where it would be difficult to become a producer of chemicals already made by established companies who controlled their technologies fiercely and were unwilling to open doors for new entrants through licensing

---

[7]Spitz. op.cit. 201.

arrangements. While the chemical world had paused its oligopolistic policies to lend its support to the conduct of the war, there was no reason to believe that the management of domestic chemical companies would not resume the conduct of their industry in the clubby ways under which it had operated only a few years earlier But the advent of new technologies based on hydrocarbon feedstocks created a situation where the industry was about to undergo tumultuous change. How this occurred is discussed in the following chapters.

2.1.12   In Europe: Little Interest in Petrochemicals

The German chemical industry in the 1930s was more advanced than chemical industries in other Western countries. However, none of the European countries had a petroleum or natural gas industry to speak of, though they, of course, did have oil refineries. In France, natural gas was discovered in 1950 at Lacq in the Southwest of the country, but there was no gas production until 1958. Italy's geologists found some natural gas in the Po Valley in the 1940s, but no chemical plants based on this gas were built for a decade or later. Great Britain's North Sea gas did not start flowing until the 1960s. Romania is probably the only European country that had found and developed substantial petroleum deposits, but it did not have much of an organic chemical industry. It is therefore safe to say that there would have been little or no interest for European chemical companies to carry out research for producing chemicals from hydrocarbon feedstocks if the issue was to replace coal or alcohol as a feedstock.

Germany: There is little evidence that I.G. Farben, which was developing high octane aviation fuel (from coal) for the Wehrmacht, was considering the fact that these gasoline blending components and byproducts would make ideal intermediates for producing organic chemicals and polymers. Germany imported all of its oil from various sources and built a number of refineries. These were using cracking processes to make gasoline and, because of the technology exchange with Exxon, received the TEL technology to raise the gasoline octane. There is no evidence to believe that German refiners were looking at refinery byproduct olefins, such as ethylene, to get into the manufacture of organic chemicals. This stems from the fact that in Germany, there was little cooperation between refineries and chemical manufacturers, with I.G. Farben companies such as BASF, Bayer, Hoechst and Huels dominating the chemical arena.

All of this is to say that the development of hydrocarbon-based chemicals proceeded quite differently in Germany—much later than in the U.S., since Germany in the 1930s was preparing for war based almost totally on coal as a raw material for both fuels and chemicals: the supply of oil that had previously been imported for Germany's refineries was almost completely cut off by the eventual successful interception of oil supply by sea, which had been anticipated by Hitler. Thus, Germany's war effort as well as its need for domestic fuel consumption was almost completely based on coal.

German scientists in the 1930s carried out substantial research on polymers such as polystyrene, polyethylene and nylon, but there was little construction of plants to make these materials in large quantities. The contrast with what was happening in the United States during the same period is hard to overemphasize.

Great Britain: Several chemical firms had, in 1926, been consolidated into was became known as Imperial Chemical Industries, producing explosives, organic chemicals, chlor-alkali and a number of other products. ICI produced acetylene from calcium carbide and separated and refined coke oven-based chemicals similar to what was done in the U.S. and in other European countries. As a charter member of the IGFarben cartel, ICI produced a number of chemicals controlled by the cartel and had access to some German technologies, notably hydrogenation of heavy tars and oils, which it practiced at its Billingham site. Chemicals were also produced from ethyl alcohol.

ICI, between the two world wars, had formed a strong alliance with DuPont, with product and technology exchanges, particularly in the area of paints and coatings.

ICI's prominent role in the discovery and commercialization of high pressure polyethylene is covered elsewhere in this book.

France: The prewar French Chemical industry was highly diversified and scattered, including a number of companies created and still headed by individuals. Important companies that wholly or in part produced chemicals included Rhone Poulenc, St. Goblin, Pechiney, and Kuhlmann, the latter primarily a chemicals producer which in the late 1920s became associated with coal companies in the Northeast of France. Coke oven-based ethylene was converted to ethylene oxide and glycol using chlorohydrin technology. Pechiney (Compagnie des Produits Chimiques et Electrometallurgiques) operated electric furnaces to make aluminum and ferro alloys and also produced acetylene and chlor-alkali, which it converted to chlorinated solvents. Société Huiles, Goudrons et Derivés (HGD), founded in the 1920s, was established to process much of the coke oven chemicals and tar in the Lorraine region, while an associated company undertook the separation of benzene, toluene and xylene, as well as phenol.[8]

Given the lack of any significant leadership on the manufacture of chemicals, particularly organic chemicals, in France, it becomes obvious why little or no research was done that was in any way comparable to what was occurring in the U.S. in the period just before the war. It is fair to say that the country was much more focused on the inorganic sector, dyestuffs, manmade fibers and specialty chemicals than organic chemicals and polymers suited for plastics and synthetic fibers.

2.1.13 Chemical Engineering: The Enabler of an Industry

While the profession of Chemical Engineering blossomed in the United States in the 1920s and 1930s, history tells that it actually goes back to the late 1980s in England. George E. Davis, who studied at the Royal School of Mines (now part of

---

[8]Fred Aftalion (1991).

Imperial College, London) worked on soda manufacture and was a consultant to the chemical industry and inventor of 67 patents. He wrote an influential *Handbook of Chemical Engineering* and identified chemical engineering as a discipline. While his lectures were criticized as being "commonplace know-how", defining practices in use by British chemical companies, he initiated new thinking in the U.S. chemical industry and this led to greater interest in chemical engineering degrees at several U.S. universities.

In 1903, five students at the University of Michigan were the first graduates with a degree in chemical engineering and by 1908 this emerging program was listed separately in the university's College of Engineering announcement. In 1912, ninety-eight percent of all patent applications registered at the U.S. patent office related to chemicals belonged to German inventors. So, after the United States helped the Allies to defeat Germany in the First World War, U.S. chemical executives with the rank of officers seized all German patents and made them available to American companies. This opened up big opportunities for chemical-related industries and spurred enrollment in chemical engineering courses that were already being offered around that time.

To understand the important role of chemical engineers in creating a chemical industry involving continuous processing and the development of process technologies after discovery of new reactions, it is useful to quote from a book on the global chemistry in the age of the petrochemical revolution[9]:

> The chemical industry was successful ... because it developed the ability to apply chemistry and chemical engineering to the manufacture of new compounds and materials ... chemists became very clever at making new molecules, but the process of determining efficacious uses of those molecules still required an unusual degree of insight ... recognition that the new compound has a combination of unusual properties that might find commercial applications ... (requiring) familiarity with far-flung networks of chemical users across a broad spectrum of industries. And using it to guide research and development.

> *After a laboratory breakthrough has been made and a potential use identified, developing a process and scaling it up required other skills. Controlling chemical reactions was one major challenge ... even if a chemist could make a reaction go the way he wanted in the laboratory, the scaling up to a large process involved numerous technical problems that had to be solved empirically.*

Arthur D. Little, (often considered as the founder of chemical engineering) *articulated a vision of the chemical industry consisting of three principles. (a) basic chemical principles, such as the conservation of matter and standard techniques, such as analysis, (b) The fact that the natural chemicals that are used in processing industries could be made synthetically or replaced entirely with synthetic ones and (c) that actual materials could also be replaced by new or synthetic ones.*

Little, at the Massachusetts Institute of Technology (MIT) early in the Twentieth Century, stated that chemical engineers concerned themselves with the *design and analysis* of *processes* as a distinct approach when considering how to make a chemical product. This led to the concept of *unit operations* as a term that emphasized the underlying unity among seemingly different operations. The operation that separates

---

[9]Louis (2007).

Fig. 2.1 Arthur D. Little. *Source* Science History Institute

water from alcohol is the same as the one that separates different fractions of crude oil from each other, namely the generation of a vapor that has a composition different from that of the liquid being distilled. There also was the understanding of a concept called *unit processes*, describing reactive processes such as oxidations, alkylations, dehydrogenations, etc. Up to the time that Little showed how to look at chemical processes differently than before, students learned everything they should know about the manufacture of, say, sulfuric acid or of paint. What students should learn, Little said, was that all chemical processes consisted of different combinations of a small number of operations such as distilling, cooling, heating, crystallizing and drying. By understanding these operations, chemical engineers could apply their knowledge to the manufacture of any chemical (Fig. 2.1).[10]

The unit operations concept sharply delineated the domain of chemical engineering and distinguished it from industrial or applied chemistry and from mechanical engineering. Industrial chemistry focused on products, applied chemistry, on the indi-

---

[10]Pap A. Ndiaya op.cit. 34.

vidual reactions employed in manufacturing and mechanical engineering on machinery. None had a focus on *processes* or recognized the operations common to a variety of products, reactions and machinery. The concept of unit operations established an independence for the fledgling profession of chemical engineering.[11]

William H. Walker, also at MIT at that time, joined Little in developing the idea of *unit operations* while helping to create a department that sponsored a research laboratory and a school of *Chemical Engineering Practice*. Separating Chemical Engineering from the Department of Chemistry was an important step. Warren K. Lewis, who had graduated from MIT's chemical engineering program after earning a PhD in organic chemistry from the University of Breslau in Germany, joined the teaching staff at MIT in 1908. Lewis led the faculty in figuring out how best to teach students the principles of *unit operations*. He used what some called a "bombastic" style, though he was beloved by students. He was shortly chosen to head up the new chemical engineering department. In 1920. Walker and Lewis were joined by Dr. William H. McAdams and Dr. Edwin R. Gilliland in publishing a landmark book on the profession (*Principles of Chemical Engineering McGraw Hill 1923*). This book became the "bible" at many other universities and colleges that offered courses in chemical engineering. Many considered Lewis the father of the profession. In an article published in 1976, Robert L. Pigford, an eminent professor of Chemical Engineering at the University of Delaware wrote, "Not only did he guide the development of *unit operations* instruction at MIT but, possibly more important, he insisted that chemistry not be neglected. In his book "Industrial Stoichiometry" he showed how simple material and energy balances could lead to insight into a process. He helped the students of his day to see that the physical nature of unit operations had to be paired with applied chemistry to solve some of the more interesting problems. Lewis's teaching led to an understanding that chemical engineers must be disciplined in the calculation of kinetics and the heat and mass transfer inherent in the design of processes and equipment. They must work with organic chemists doing heterogeneous catalyst research and use their training to help in planning experiments and in the interpretation of results.[12]

Lewis visited German chemical firms, recognizing that country's preeminence in chemicals manufacture. He observed that while Germany industry was well ahead of the U.S. in developing organic chemicals manufacturing, its approach of having chemists working with mechanical engineers resulted in a lack of basic understanding of how processes actually work. When studying the German way of developing technology, he saw that details of equipment construction were left to mechanical engineers, and that those designers were implementing the ideas of the chemists with little or no understanding of their own of the underlying reasons of how things were done. The result was a divorce of chemical and engineering personnel not only in German technical industry but also in the university and engineering schools that supplied this industry with professionally trained men. It is interesting that the term "chemical engineer" was rejected by the Germans as hybrid and unclear.

---

[11]Darton et al. (2003).

[12]Pigford. op.cit. 192.

Writing in the above-quoted article, Pigford also wrote about *transport phenomena, which started being regarded as one of the engineering sciences that are basic to understanding engineering problems and that are highly mathematical analyses of phenomena on a small scale. Thus, equipment design is suppressed and the unit operations have coalesced into transport of energy, mass and momentum … our thought has been that the subject of transport phenomena should rank along with thermodynamics, mechanics and electromagnetism as one of the key engineering sciences.*[13]

At DuPont, Thomas H. Chilton, head of the Department of Chemical Research and Allan P. Colburn spearheaded a new Chemical Engineering department, developing a more generalized, conceptual approach to the discipline. Chilton urged chemical engineers to adopt four steps for new process development, (1) initial identification of alternative processing steps to be considered, (2) laboratory studies of the reactions involved and selection or design of the equipment, (3) pilot scale operation and (4) design of the initial plant unit. Colburn showed that some of the central chemical engineering operations, such as momentum fluid flow, heat transfer and distillation could be understood as analogous phenomena. T. K. Sherwood, another important figure in advancing the principles of chemical engineering, stressed that experimental data should be the primary source of information rather than mathematical analysis, when applied to mass transfer problems. Colburn won the AIChE's first William H. Walker award for an outstanding contribution to chemical engineering literature. Crawford H. Greenewalt, later president of DuPont, who received a chemical engineering degree at MIT, studied under Lewis and brought the "scientific principles of *unit operations*" to DuPont's chemical department.

DuPont was very much engaged in the development of chemical engineering at MIT. The company was particularly interested in MIT's Research Laboratory of Allied Chemistry, which enrolled the most talented undergraduate and graduate students. DuPont's main objective was to recruit bright chemical engineering students through financial support, the submission of research topics and the hiring of consultants. The company established several fellowships and later also provided direct financial support to the school, when the latter was in some financial difficulties. Dr. Karl Compton, who headed up the university, appealed to captains of industry for support, not only to DuPont and its executives like Lammot and Pierre DuPont, but to General Electric, General Motors and Eastman Kodak, among others, who came through with funds for MIT. A number of research topics, some suggested by DuPont, were carried out as a result of these contributions, notably research on gases at high pressures. This research was of high priority to DuPont and led to its large scale production of ammonia and nitric acid.

DuPont also became very interested in polymers and hired two consultants, Roger Adams and Carl "Speed" Marvel from the University of Illinois to help DuPont in entering this field. Marvel's knowledge of Dr. Wallace Carothers, soon to become the inventor of nylon, undoubtedly steered that famous chemist to DuPont.

---

[13]Ibid, 182.

Chemical engineers became more and more important to DuPont as a result of the advances made in the field of *unit operations*, which gave the chemical engineering profession a sense of specialization. The chemical engineers who joined DuPont in the 1920s and thereafter were "the specifically American products of a carefully calibrated course of technical studies designed to respond to the new needs of industry."[14]

DuPont's conviction regarding the importance of chemical engineering as a profession uniquely capable of guiding the future of DuPont and capable of leading to major industrial successes was the basis of Chilton's speech when he received the Chandler Medal in 1939. Rewarded for his remarkable merits in the discovery and the formulation of the principles underlying *unit operations* in chemical engineering, and for the application of these principles to the development and the design of chemical plants he said "*the commercialization of these synthetic materials … is not miraculous, but is the result of painstaking application of sound engineering, as well as of brilliant chemical research*".[15]

The number of students taking Chemical Engineering as a major course rose from 869 in 1910 to 5743 in 1920 and 12,550 in 1936. One author, citing the rising interest in this profession, stated that "One academic institution, MIT, played a decisive role in its rise to power."[16] The university was the first to award a PhD in Chemical Engineering in 1923. Another key development, led by Warren K. Lewis and Robert T. Haslam, was the establishment, unique for its time, of a Chemical Engineering Practice School with stations at industrial production sites (Initially, a paper company, a steel company and a detergent factory.). Engineers assigned to work at these locations under the supervision of an MIT faculty member, engaged in problem solving, using *unit operations* methods as taught at MIT.[17]

Lewis had helped oil refiners in increasing the yield of gasoline from crude oil. Notably, he has been credited for the successful application of the scientific principles of fractionating column design to the oil industry. Lewis crowned his career heading up the development, in the early 1940s, of the fluid bed catalytic cracking process.[18] This novel technology converted, on a continuous basis, heavy hydrocarbon oils to gasoline range fractions by pairing a reactor, where carbon was laid down on the catalyst, with a regenerator, where the carbon was burned off and the "regenerated" catalyst returned to the reactor. This was obviously superior to the then employed Houdry cracker, where a number of fixed bed reactors had to be taken off line periodically and regenerated in a cyclical manner. The idea of carrying out a reaction by suspending fine particles in a rising current in a bubbling bed evidently came from Lewis, who termed it "jiggling". Lewis claimed that the transport of heat in the bubbling bed of a fluidized reactor was extremely fast, providing uniformity

---

[14]Pap A. Ndiaya op.cit. 56–62.

[15]Ibid 105.

[16]Ibid 49.

[17]Furter (1980).

[18]Rosenberg et al. (1992).

of reaction temperature.[19] Exxon, which had a close relationship with MIT, sponsored the development work on the fluid bed cracker just before the start of the war. The Model IV catalytic cracker, an evolved design, became the standard of the petroleum industry. Fluidized bed processing later also became a technology used in petrochemical processing, such as the production of acrylonitrile from propylene, ethylene oxychlorination and propane dehydrogenation.

DuPont and Exxon were, of course, not the only companies that forged strong alliances with chemical engineering departments at universities. This kind of link and the dependency of universities to depend on industry funding provided a strong impetus for universities to focus on industry needs. There was a constant flow of chemical and chemical engineering talent between universities and companies. To avoid losing experienced talent to universities, companies frequently adapted their employment conditions to those found in universities, allowing professors to maintain a degree of freedom and flexibility and, at times, to publish the results of their research.[20]

By the late 1930s, MIT was graduating a substantial number of chemical engineers, many with advanced degrees. With faculty members such as Drs. Lewis and Gilliland having contacts and consulting assignments with firms such as Exxon and DuPont, MIT chemical engineering students were exposed to industrial developments involving both petroleum and chemical processes, which were becoming more complex. The important advances in catalytic processes, such as the conversion of previously largely useless (except as fuels) olefins to polymer gasoline and alcohols undoubtedly excited students interested in novel technologies that could be applied in other areas (Fig. 2.2).

In his oral history, James Fair, later a top petrochemical executive at Monsanto, recalled that "chemical engineering in the 1930s was thought to be a glamorous field with great opportunities. It was, particularly so, as it turned out, with the war coming along so suddenly."[21] Fair had transferred from The Citadel to Georgia Tech, which had a chemical engineering department. He remembered that between his junior and senior year there, students came back in the summer to design and build a pilot plant for the production of aniline from benzene and nitric acid. It was a continuous process and would be considered a petrochemical one if the benzene were to come from refinery operations rather than from coke ovens. Fair mentioned that many of the most illustrious chemical engineers he met or was aware of during his career, all involved in the petrochemical industry, had done their graduate work in the 1930s.

With chemical companies still producing organic chemicals, plastics and man-made fibers from coal, alcohol or wood, but with research pointing the way toward use of hydrocarbon feedstocks and with more and more chemical engineers being trained to use new methods for the design of complex catalytic processes, the stage was set for a new era.

It is clear why a number of MIT chemical engineering graduates were attracted to Scientific Design Company, which would give them an opportunity to apply the

---

[19]Pigford, op.cit. 194.
[20]Galambos. op. cit. 36–37.
[21]Fair (1992).

Fig. 2.2 Professor W. K. Lewis speaking to students at the MIT Chemical Engineering Practice School station at Exxon's Bayway, N.J. refinery ca. 1942

principles they had been studying in the university in an atmosphere of novelty and excitement. In the late 1940s, MIT students earning a Master's degree in Chemical Engineering Practice returned to school after solving problems at the three Practice School stations to spend a month as a team to practice designing a chemical plant. The author recalled that his group, which had attended Practice School stations at a paper factory, a steel plant and an explosives and plastics manufacturing plant, was given the assignment of designing a polystyrene plant. At that time, Dow, Monsanto and others were already producing styrene as part of the synthetic rubber program producing styrene-butadiene rubber. The students produced a flowsheet with heat and material balances and equipment designs that included reacting benzene with ethylene over a catalyst (alkylation), separating the unreacted benzene from ethylbenzene (distillation), separating ethylbenzene from reaction byproducts (distillation), and cracking the ethylbenzene over a catalyst at very high temperatures to styrene (dehydrogenation and more distillations) to produce highly pure styrene. This was then reacted with a catalyst under relatively mild conditions to form polystyrene (polymerization of a liquid to a solid), followed by cooling and pelletizing.

When students with this type of background entered operating companies or research and engineering firms, they were eager to apply their knowledge of unit operations to develop new processes using reactive hydrocarbons such as olefins, benzene, toluene and xylene. Scientific Design, which hired a number of MIT graduates, initially decided to focus primarily on alkylations and oxidations, with exceptional results.

References

Darton, R.C., R.G.H. Prince and D.G. Wood (2003) *Chemical Engineering: Visions of the World.* ELSEVIER

Fred Aftalion (1991) *A History of the International Chemical Industry* University of Pennsylvania Press, Philadelphia, Pa. 281

Galambos, Louis. (2007) *The Global Chemical Industry in the Age of the Petrochemical Revolution.* Cambridge University Press. 170–171

Graham, Edgar (1939) *Tetraethyl manufacture and use.* Industrial and Engineering Chemistry Vol 31, No. 12. 1439–1446

James Fair (1992) Oral History Program, Chemical Heritage Foundation, Philadelphia Pa

Peter H. Spitz (1988) *Petrochemicals. The rise of an industry* John Wiley & Sons New York. 125

Robert L. Pigford (1976) *Chemical Technology: the past 100 years.* Chemical & Engineering News. April 6, 194

Rosenberg, Nathan. Ralph Landau and David C. Mowery (1992) Technology and the Wealth of Nations. Stanford University Press. Stanford, Ca. 91

William F. Furter. (1980) *History of Chemical Engineering* American Chemical Society, Washington, DC. 85

Bibliography

Borkin, Joseph (1978) *The Crime and Punishment of I.G. Farben.* Free Press N.Y.

Haynes, Williams. (1942) *The Chemical Age.* Alfred A. Knopf. New York

Hougen G.A. (1935) *Industrial Chemical Calculations.* John Wiley & Sons. New York

Lewis, Warren K. (1953) *Chemical Engineering – A New Science?* Centennial of Engineering 1852–1958 Chicago Museum of Science and Industry

Scriven I.E. (1991) *On the Emergence and Evolution of Chemical Engineering. In Perspectives in Chemical Engineering* C.K. Colson. Ed. Academic Press

Sherwood, Thomas K. (1937). *Absorption and Extraction.* McGraw Hill and Co., New York

Walker, William H., Warren K. Lewis, William H. McAdams and Edwin R. Gilliland. (1937) *Principles of Chemical Engineering.* McGraw Hill Book Company, Inc. New York

Whitehead, Don. (1968) *The Dow Story.* McGraw Hill Book Company, Inc. New York

Chapter 3
Fuels and Chemicals Research Helps Win World War II

Abstract Research on high octane fuels carried out by some oil companies and UOP also branched out into research that could produce chemicals from petroleum fractions and natural gas liquids. Union Carbide, Exxon, Shell, and Dow became pioneers in early petrochemical R&D. All of this research proves immensely valuable as the U.S. is drawn into World War II and the synthetic rubber and aviation gasoline programs. The important relationship between Exxon and I.G. Farben is discussed. Commercialization of polyethylene, polystyrene and PVC as well as nylon and polyester fibers occurs just before or during the war.

In the 1930s, researchers and chemical engineers working on basic organic chemistry in the United States were often employed at large companies producing chemicals from coal, alcohol, or wood. A few firms were, however, starting to look at hydrocarbon feedstocks, given the abundance of petroleum fractions and natural gas and gas liquids in certain parts of the country, notably the U.S. Gulf Coast, Oklahoma and California. At a few oil companies laboratories, there was also interest in the use of aliphatic feedstocks, but their focus was mainly on the fuels industry as automobiles and military aircraft started to use higher performance engines. These required gasoline components that would keep engines from knocking, such as branched aliphatics and BTX (benzene, toluene, xylene) aromatics. While these high value gasoline components were being separated and recovered from petroleum refinery cracking operations, more and more so-called high octane components were required every year and so research was conducted to either synthesize such molecules or to convert naphthas with novel technology to higher octane blending components. This area of research was stepped up as World War II was approaching and the need for large amounts of high octane aviation gasoline became critical. Universal Oil Products (UOP), a firm created by a number of smaller oil companies as an independent research organization to develop new refining technologies, contributed to the development of high octane gasoline components. In 1932, it announced the production of a hydrocarbon alkylate in the naphtha range by reacting propylene and butylenes over a solid phosphoric acid catalyst. This so-called polymer gasoline considerably improved engine performance. The same alkylation technology was a few years later used to propylate benzene to make cumene (isopropyl benzene), which became a very

© Springer Nature Switzerland AG 2019
P. H. Spitz, *Primed for Success: The Story of Scientific Design Company*,
https://doi.org/10.1007/978-3-030-12314-7_3

high octane gasoline component. German chemists were also working diligently and under great pressure to develop technology to produce high octane aviation fuels, as well as diesel fuel for military vehicles. Essentially all of this research was based on coal or lignite feedstock, since only a very limited amounts of crude oil and natural gas were produced in Germany.

An important development in Germany chemistry was the electric arc process for making acetylene from domestic natural gas, with the first plant built by Huels in 1940. This could be considered as the first "petrochemical" plant in Germany. (i.e. production of a reactive hydrocarbon molecule not derived from coal or alcohol.) The so-called Wulff process for acetylene was employed to build several plants in Europe and in the U.S. after the war. However, acetylene was eventually replaced with (less expensive) ethylene for vinyl chloride and vinyl acetate production, among other uses.

Both in Germany and in the U.S., chemists started to look closely at polymers as a class of materials. In Germany in the 1920s, Hermann Staudinger hypothesized that certain polymers are composed of smaller materials, including monomers, linked together in head-to-tail fashion like paper clips, including repeating molecular units. X-Ray diffraction studies in the U.S. by Hermann Mark supported this theory. Staudinger further proposed that rubber, starch, cellulose and proteins are long chains of repeating materials. Staudinger's work earned him the Nobel Prize.

Celluloid, one of the earliest "plastics", consists of chains of cellulose, as does rayon, cellulose fiber, nitrocellulose and cellulose acetates. In the 1920s, thermosetting polymers (e.g. phenolic or urea formaldehyde resins, often referred to as Bakelite,) became important materials also known as "plastics" and found many uses such as telephone headsets, billiard balls, dinner dishes and electrical components. Polymers of this type cannot be melted once they are heated and solidified and became known as thermosetting resins. This is a serious limitation in processing, which led to research in the 1930s on so-called thermoplastic polymers. Thermoset resins were the only plastics in common use before the start of World War II (Figs. 3.1 and 3.2).

Fig. 3.1 Cellulite objects. *Source* Science History Institute. Jane E. Boyd/George Tobias

Fig. 3.2   Phenolic and acetate buttons. *Source* Hilda Spitz

An important direction of polymer research in the 1930s was concentration on so-called thermoplastic polymers, characterized by the fact that they are produced in pellets that can be melted and shaped by heating and used to produce various parts and objects by techniques like injection molding, extrusion and thermoforming. Plastic processors greatly prefer thermoplastic polymers due to the versatility of their use. Significantly, polystyrene, and polyvinyl chloride, all more economically produced from hydrocarbon feedstocks, were commercialized in small plants by companies like Dow, Monsanto and B. F. Goodrich in the 1930s and became polymers of choice in the petrochemical era.

German chemists and mechanical engineers had succeeded in the early part of the century in synthesizing ammonia from nitrogen and hydrogen in a high pressure reactor. This also started research on other chemical reactions at very high pressures, largely carried out in European countries, including the Netherlands. At Leiden, a cooperative study with ICI led to the successful synthesis of low density polyethylene. This polymer was commercialized just before the start of World War II.

In Germany, England and the U.S., research was also being conducted to develop technology for the production of synthetic (versus man-made) fibers, notably polyamides (principally nylon) and polyester. Dr. Wallace Carothers' work at DuPont stands out as the most important breakthrough in this area of research. DuPont's outstanding work in synthetic fibers, which later led to acrylic fibers (Orlon), was based on the company's desire to produce fibers that would overcome the shortcomings of rayon and silk, such as silk's yellowing with exposure to sunlight and poor resistance to chemicals.

In short, the fruits of research carried out in the 1930s and early 1940s culminated in the commercialization of a large number of new processes and technologies over a short period just ahead of the war. This chapter will describe in more detail some of the most important areas of research, which helped the Allies win the war and later became the backbone of the new petrochemical industry.

In this book there are descriptions of a number of chemical "processes" and it therefore may be useful to define what is meant by that term. *A process is fundamentally a method of carrying out a manufacturing sequence.* In the chemical industry, processes include changes of state or chemical reactions or both. They require a certain amount of experimental work, often resulting in successful commercialization. This would normally lead to a patent, and to the development of technical information, often referred to as "proprietary knowhow".

In the 1930s, a limited number of companies started to develop processes based on hydrocarbon feedstocks. This was the birth of the petrochemical industry.

3.1 Companies Start Looking at Hydrocarbon Feedstocks

3.1.1 Union Carbide

This story starts with the formation, in 1904, of Prest-O-Lite Company, a distributor of acetylene gas sold nation-wide in cylinders for lighting and welding. The firm shortly became the largest purchaser of calcium carbide for acetylene production and of acetone in which the acetylene was dissolved under pressure. Union Carbide was then the largest producer of calcium carbide (for electrodes) and the Presto-O-Lite company its largest customer.[1] Considering the desirability of integrating backward, Presto-O-Lite decided to sponsor research on acetylene and later ethylene at Mellon University in Pittsburgh. Aware of this, Union Carbide then also hired Mellon, but to work on making acetylene and ethylene derivatives from natural gas. This work was conducted under the supervision of George O. Curme, who oversaw the cracking of propane and higher hydrocarbons to acetylene and ethylene using an electric arc. Union Carbide was also aware that the U.S. arm of Linde, a German oxygen producer, was working on a way to make ethylene from natural gas, their interest coming from the fact that separating the reaction products of ethane cracking required refrigeration (cryogenic) technology with which Linde was very familiar due to its air separation business. Linde produced the ethylene by cracking ethane in an electrically heated silicon tube.[2] This may be the first time that ethylene was produced on purpose by cracking a hydrocarbon in a heated tube.

In 1917, the three companies decided to throw their lot together, with emphasis on developing a commercial process to make ethylene from natural gas. A new company was founded, called Union Carbide and Carbon Company, which later became known as Union Carbide. Research was led by Curme, who later received the Perkin prize for the inventions he spearheaded during this period.[3]

---

[1] Spitz (1988).

[2] J.N. Compton (1937) *Informal Personal Observations on the history of the Carbide and Carbon Chemical Company.*

[3] G. O. Curme, Jr. *Industrial Toolmaker.* Industrial & Engineering Chemistry 27. February 232–230.

In 1920, the company bought a small refinery near Clendenin, West Virginia to experiment with promising research from Mellon.[4] Linde's work on ethane cracking soon led to successful production of ethylene in a tubular furnace. A small plant was constructed to make ethylene in this manner, which came into operation in the early 1920s. A subsidiary was then created to develop and commercialize the production of a number of aliphatic chemicals previously produced either from acetylene or from ethyl alcohol. Earlier, Mellon had discovered that ethylene glycol could lower the freezing point of a solution of glycol and water and could therefore be usefully employed as automobile antifreeze. Ethylene from the new plant was converted to ethylene oxide via the chlorohydrin process and the oxide hydrolyzed to ethylene glycol. Union Carbide started to market this material under the name of *Prestone*.

Union Carbide soon thereafter learned how to crack propane and butane to make propylene and butylenes, useful for further processing into various derivatives. Then, a larger plant was built and started up in 1926.[5] In 1937, Union Carbide built the first plant making ethylene oxide via direct oxidation of ethylene using a silver catalyst. Work on ethylene derivatives also resulted in the production of other ethylene derivatives, including synthetic ethyl alcohol, ethylene glycol, ethyl ether, triethanolamine, vinyl chloride, vinyl acetate and others. Some propylene derivatives also started being made. By 1934, Union Carbide made 34 derivatives from ethylene and 15 from propylene.[6]

In retrospect, one might wonder why some other chemical companies did not at that time follow Union Carbide into ethylene production via hydrocarbon cracking… The answer, most probably, is that the major ethylene derivatives: polyethylene, polystyrene, vinyl chloride, ethylene oxide and glycol only became multimillion pound chemicals after the war. Without these large markets, ethylene was mainly used for relatively low volume organic chemicals, a market that Union Carbide was dominating. Exxon in Baton Rouge was cracking propane to make ethylene in the thirties, but only in connection with producing ethyl chloride for TEL.

3.1.2 Shell Chemical

Shell Oil became aware in the late 1920s that Germany was developing technology that would convert its abundant coal into liquid fuels and chemicals. The so-called Fischer-Tropsch process made chemicals, as well as diesel fuel and other hydrocarbons. Seeing this as a threat to its business, Shell decided it should figure out how to make chemicals from petroleum products as a sort of counter to the German initiative. Europe had little indigenous natural gas and gas liquids so it was decided to commence chemicals research in Emeryville, California, where Shell had a refinery.

---

[4]The West Virginia Encyclopedia.

[5]Robert L. Pig ford (1976) *Chemical Technology: the past 100 years. Chemical & Engineering News,* April 6, *1976.*

[6]*P. H. Spitz* op. cit. 81.

Shell's first foray into petrochemicals was actually based on its decision to license the St. Denis process for ammonia manufacture, which was outside the scope of the Haber-Bosch patents assigned to I.G. Farben. Shell purchased two ammonia plants in Europe and erected them in California. While the company did not achieve the desired commercial success with ammonia prices falling over this period, the experience proved valuable to the Shell organization and other firms also started to look at ammonia as a use for natural gas. A number of ammonia plants were built during the Second World War to support the production of nitrogen-based explosives and artillery shells, as well as for fertilizers. Produced increasingly from natural gas to augment the amounts recovered from coke ovens, ammonia became part of the petrochemical industry, soon converted into nitric acid for nylon production, acrylonitrile for acrylic polymers (e.g. Orlon) and acrylonitrile-butadiene-styrene resins.

The first product coming out of hydrocarbon-based research in Emeryville was actually a chemical, namely secondary butyl alcohol, produced from n-butylene and iso-butylene. The next step was dehydrogenating the alcohol to methyl ethyl ketone (MEK) which turned into a successful commercial operation, given that MEK was a solvent for most organic materials.

The researchers also turned their attention to the development of high octane gasoline additives and here they were also successful. Having isolated isobutylenes for MEK, it was found that dimerizing this chemical to di-isobutyl, followed by hydrogenation, resulted in isooctane, a key gasoline additive and, when combined with tetraethyl lead (TEL), a very important aviation fuel. Shell also started to produce isopropyl alcohol and acetone from propylene.

The most complex chemical synthesis discovered by Shell was the production of glycerin. It involved chlorination of propylene to allyl chloride, which could be converted to allyl alcohol and glycerin. Epichlorhydrin, a glycerin intermediate, decades later became the raw material for epoxy resins.

An important Shell development for the war was an extractive distillation process developed in 1940 to recover toluene, again a high octane gasoline blending component. Twenty-eight plants were built during the war, using this process, by Defense Plant Corporation.

3.1.3   Standard Oil (N.J.) (Exxon)

In the late 1920s, Exxon became the first oil company to make a "petrochemical" from an olefin, constructing a plant to convert refinery propylene to isopropyl alcohol. It used sulphuric acid to make isopropyl hydrogen sulphate which was hydrolyzed to the alcohol.

Around the same time, Exxon became aware of the high pressure hydrogenation work being carried out by I.G. Farben on coal and entered into a partnership with the German organization. This turned out to be a crucial relationship that greatly facilitated the efforts of both the U.S. and Germany to develop novel fuels and chemical technologies and is described in detail in the next section. The German firm was already working on synthetic rubber development and included this area in the agreement. Information gained from I.G. Farben opened up the whole area of catalytic processes for Exxon, not only for refining but also for chemical applications.

Ethyl Corporation started producing tetraethyl lead (TEL) in 1922 at Exxon's Bayway refinery in New Jersey, using ethyl chloride as a feedstock. The ethylene required was recovered as a byproduct from oil cracking operations and this was soon found to be insufficient as TEL demand was increasing rapidly. This led to the additional use of ethyl alcohol dehydration as a source of ethylene. Some years later' Exxon, similar to Union Carbide, started to look at ways to produce ethylene on purpose from hydrocarbons, specifically by cracking propane. Such a plant was built and successfully operated at Exxon's Baton Rouge, Louisiana refinery, the first cracker built by an oil company in the United States. It was turned over to Ethyl Corporation in the early 1930s. Exxon continued to experiment with hydrocarbon cracking to ethylene, including the use of heavier feedstocks such as naphtha. This development work was also done at Baton Rouge. While Union Carbide had pioneered light hydrocarbon cracking, Exxon was the first to crack naphtha and gas oil successfully.

With military uses of high powered gasoline increasing, Exxon turned its attention to producing high octane blending materials, similar to what Shell had been developing. Exxon found that butylene and isobutene could be alkylated to a dimer with sulfuric acid to make a valuable high octane blending stock. Isobutylene could be dimerized and hydrogenated to isooctane. Exxon's plant at Baton Rouge became the Allies' largest source of high octane aviation fuel and was said to have saved England in the Battle of Britain.[7]

In the mid-1930s, Exxon's and Standard Oil Co. (Ind.) researchers experimented with the hydrogenation of naphtha, again turning to German catalytic knowhow. They found that using a base metal catalyst at considerably elevated temperatures and pressures, and hydrogen, low octane naphthas could be converted to much higher octane values when the straight chain aliphatic hydrocarbons were converted to branched chain molecules and aromatics. The engineering development of this process, was done at M.W.Kellogg. Interestingly, this was one of Ralph Landau's assignments at that firm in 1941 (see Chap. 3). Exxon was encouraged by the government's Ordnance Department to commercialize this process, which came to be known as Hydroforming. In 1940, 20,000 gallons of synthetic toluene were produced in available equipment. A large plant was built at Bayway, which supplied toluene for aviation fuel during the war.

---

[7]Gornowski (1980).

3.1.4 Dow Chemical

Dow, headquartered in Midland, Michigan was already a large chemicals company before World War II, producing chlor-alkali, bromine products, vinylidene chloride (Saran) and phenol, among many other products. The company in 1922 pioneered U.S. production of synthetic phenol from benzene and chlorine under high pressure, but demand for this product was low due to the large supply of phenol from coke ovens. Dow's interest in ethylene was linked to its TEL joint venture with Ethyl Corporation, where ethyl bromide was added to TEL to eliminate deposits on engine valves, Dow being the largest producer of bromine from wells in Michigan. Work on styrene synthesis was started in the late 1920s but was discontinued for a while.

Looking for another source of bromine, as Midland's brines were starting to become depleted, Dow decided to go to sea water as a more plentiful source, which would also make Dow a producer of magnesium from sea water. A Texas location would favor the production of ethylene from abundant natural gas liquids on that coast and electricity was cheap. Dow had begun to produce ethylene oxide and other ethylene derivatives at the Midland plant and was looking to expand. Accordingly, Dow built what turned out to be the first industrial chemical complex on the Gulf Coast that, in addition to bromine and magnesium salts and metal, would produce 25 million pounds of ethylene and 15 million pounds(each) of ethylene glycol, ethylene dichloride, ethylene dibromide and vinylidene chloride.[8] It started up in 1941. Around this time, the company had become interested in making styrene and polystyrene and these plants were added shortly afterwards, with styrene built at Velasco, Texas not far from Freeport.

A major contribution by Dow was the open construction of plants, different from the chemical industry's tradition of putting chemical operations inside buildings, as was always the case in Germany and elsewhere in Europe as well as in the United States by companies like DuPont and Monsanto. After Freeport, most other plants on the Gulf Coast and in warm climates in general used open construction. Even in colder climates, few firms went back to the traditional manner, given the fact that open construction tends to be a third less expensive than putting operations inside buildings.

3.1.5 Other Companies

The first methanol plants based on natural gas were built in 1927 and 1929, respectively, by Commercial Solvents Corporation and Cities Service Company, both producing less than 10 million gallons per year. DuPont had built a somewhat larger gas-based methanol plant in Belle, West Virginia near its ammonia plants, but

---

[8]P. H. Spitz op.cit. 95.

it was based on gasifying coke. Methanol eventually became one of the highest volume petrochemicals, its main use being for the production of acetic acid and formaldehyde.

3.2 Petrochemical-Based Polymers Started Their Phenomenal Growth During World War II

Polyvinyl Chloride (PVC): Discovered in the late 1800s and first used as a rigid plastic in Germany and Russia in the early 20th Century. It became more useful in the mid-1920s when B. F. Goodrich found out how to make PVC flexible by adding plasticizers. Relatively small plants were built in Germany, in the U.S. and elsewhere before World War II. In wartime, more PVC plants were built at several locations and the product was used for military tents, protection sheeting, and piping in various applications. After the war, PVC started gaining widespread use in consumer and building applications, including domestic and underground piping, shower curtains, table coverings, automobile seating, etc. Because it was made from acetylene-based vinyl chloride (VCM), PVC could not be considered a petrochemical until ethylene-based VCM started being produced.

Polyethylene: Low density, high pressure polyethylene was discovered by Reginald Gibson and Eric Fawcett of Imperial Chemical Industries (ICI). Industrial production started in England in 1939 and in Germany shortly thereafter. During World War II, in 1944, the Bakelite Company and DuPont obtained licenses from ICI and started polyethylene production in this country, totaling about 15 million pounds.[9] Union Carbide requested a sublicense from DuPont, was refused, and nevertheless proceeded to build a polyethylene plant larger that DuPont's – then got the OK for this from the U.S. Justice department as part of an antitrust settlement between ICI and DuPont.[10] Polyethylene was used for sheathing wire and cable and for radar installations. By 1949, 50 million pounds were produced and the polymer started to replace cellophane as a packaging material. Polyethylene soon became the largest volume plastic, with primary use in packaging, bottles and other domestic applications and in films for agricultural purposes.

Polystyrene: Discovered as a polymer in France and Germany in the late 1880s when it was found that heating styrene monomer produces a chain reaction that results in polystyrene. It started being manufactured by BASF in 1931. Dow Chemical built a styrene-polystyrene plant shortly before World War II and invented the Styrofoam process in 1941. Polystyrene found a number of uses during the war, but more importantly, its monomer, styrene, was a key ingredient for synthetic rubber so that a number of styrene plants were built to support this program. These could easily be expanded after the war was over and polystyrene plants built next to them.

---

[9]Popular Mechanics July 1949 125ff.

[10]Hounshell and Smith (1988).

Polystyrene's use in containers and cups and the insulating properties of polystyrene foam in housing and its use in packaging made this polymer a household name and key petrochemical.

Nylon: In the late 1927s, DuPont decided to investigate areas for promising research, including colloid chemistry, catalysis and organic chemicals.[11] The organic team included Wallace Carothers, a Ph.D. who was an instructor at Harvard before joining DuPont. Wallace soon turned his attention to the area of polymers, which had recently been elucidated by the German chemist Hermann Staudinger, who later received the Nobel Prize. The polyamide family was first identified by Carothers at DuPont's Experimental Station in 1935. It became a great success when it was found that nylon had the desired properties of elasticity and strength that silk lacked. The first nylon 6/6 plant was built in Seaford, Delaware and commenced production in late 1939. First used for tooth brushes, women's stockings and undergarments, it became the main fabric for military parachutes when it became impossible to import silk from Asia. Nylon production was greatly expanded, as parachute drops were used extensively during the latter part of the war.

German chemists were familiar with Carothers' work, which was broadly patented. However, they were able to get around DuPont's nylon patents by developing a similar polyamide, Nylon 6 based on caprolactam. This became possible, because I.G. Farben was able to show that Carothers had written that caprolactam "could never be polymerized to make a polyamide material like nylon.[12]

The first synthetic rubber produced in the United States was by DuPont, which commercialized Neoprene (2-chloro-3butadiene in 1932.[13]

3.3   German Chemistry Was Essential for U.S. War Effort

It may come as a surprise to many readers that there was strong cooperation between I.G. Farben and Exxon in the two decades before World War II. The exchange of technologies between the two parties greatly helped both to produce war materials used in the eventual, brutal conflict between Germany and the United States. The exchanges not only covered information on the manufacture of synthetic rubber, but also on the production of high octane aviation gasoline and on various chemical processes. While it was definitely a two-way agreement, it seems difficult today to understand how Exxon continued the exchange right up to the time hostilities broke out in 1939. Charitably, it may be assumed that Exxon was continuing to try and get information on Germany technology at a time when the U.S. was still trying to stay out of the European conflict. This seems unrealistic in retrospect.

---

[11] Ndiaye (2007).
[12] Hounshell op.cit. 207.
[13] Furter (1980).

In the spring of 1925, Exxon received a call from a high level executive of BASF with a request that the Germans wanted to visit several U.S. refineries. In return, BASF would welcome representatives of Exxon to Germany to observe how German chemists were starting to make gasoline from coal. This was agreed. When the Exxon representatives visited the BASF complex at Ludwigshafen they "were plunged into a world of research and development on a gigantic scale.[14] Principally, they learned that with catalysts and hydrogen, coal could be converted into high quality gasoline and diesel fuel. Hydrogenation reactors 30 feet high were operating at 3000 lb per square inch at "red hot" temperatures. It immediately occurred to the Americans that since crude oil contained only a moderate amount of gasoline, the German technology could convert heavy crude "bottoms" into more gasoline. So, the German technology raised great interest at Exxon.

The discovery and production of crude oil in the U.S. goes historically in cycles of abundance and shortages, while the country has always had a huge amount of coal reserves. Around the time of the Exxon visit to Germany, there was actually a significant drop in crude oil production in the U.S., with gasoline price rising from 40 cents to one dollar per gallon. Even if making gasoline from coal would be expensive, the technology was potentially attractive for a possible future of oil shortages. (Actually, much more oil was discovered in East Texas shortly after that time). As for Exxon's immediate interest, the Germans had not worked on the hydrogenation of heavy oils and that became the thrust of Exxon's interest at the time.

In 1926, Exxon decided to build a large laboratory in Baton Rouge, Louisiana, dedicated to exploring hydrogenation of crude oil fractions and related areas. A year later, an agreement was signed that called for an exchange of technical information between BASF (now part of I.G. Farben) and Exxon, as well as a program for Exxon to build a small plant that would produce 40,000 tons of hydrogenated oil products per year, with results to be shared with BASF. As part of the agreement, the Germans would keep all the rights for producing fuels and chemicals from coal, with reciprocal rights to Exxon for fuels and chemicals from crude oil or natural gas. Also, the I.G. agreed to offer to Exxon a minority participation in any new process I.G. developed for making chemical products from oil or natural gas. It was through this covenant that there came to America information on a synthetic rubber process. Exxon and I.G. Farben formed a company that would sell jointly-licensed technologies to third parties.

The Germans were very proud of their achievements and of the partnership, which validated German research and in addition created a steady flow of information from the I.G. laboratories in the years ahead. The two parties also joined in another company known as The Joint American Study Company (JASCO), as a vehicle for commercial testing and licensing of new processes developed by either parties for making chemicals from oil raw materials. Each new process was a separate entity, with the inventor entitled to two thirds of the rewards and the other one third.

The Buna process for making synthetic rubber from Butadiene was not originally included in the agreement, but then the partners later agreed that the acetylene initially

---

[14]Howard (1947).

used to make butadiene could be produced from natural gas. Accordingly, the Buna technology was made part of the JASCO agreement. Samples of the butadiene rubber were sent to the U.S. where several of the large tire companies conducted experiments. It was found that butadiene rubber, though useful, did not have the essential qualities needed for making good tires. In the meantime, the Americans made available to the Germans the new Neoprene process, based on chlorinated butadiene, discovered by Father Nieuland at Notre Dame. This was mainly useful for specialized rubbers, but was also under consideration for tires.

Germany's interest in synthetic rubber in 1930 was originally based on the country's large bill for rubber imports from Malaysia. Only after Hitler's concern about likely blockades of rubber shipments if war broke out, was the German chemical industry told to speed up the work and make Germany independent of rubber imports.

In 1933, the Leverkuesen Works operated by Farbenfabriken Bayer A.G. started working on a rubber-like copolymer where 25% of styrene was introduced into the butadiene chain. Within a short time, production was 25 tons per month and would reach 200 tons per month within a year. But there were considerable processing problems with this material. The I.G. had also been developing a different type of copolymer using acrylonitrile instead of styrene (called BunaN), and this at first looked promising. However, by 1938, much more work and testing on butadiene-styrene copolymer (BunaS) was starting to look good and became the basis of Germany's synthetic rubber industry. Some, but not all of this information was flowing to Exxon and the domestic rubber companies over this period. This was useful to the Germans as well, since Goodyear, Goodrich, Firestone and the other rubber companies were carrying out tests on the new synthetic rubber, with results made available to the I.G. through the JASCO agreement.

Through this agreement, the I.G. also received information about Exxon's butyl rubber research, allowing Germany to make tire inner tubes with this type of synthetic rubber.

Two other technologies were exchanged, though not through JASCO. Ethyl Corporation made the tetraethyl lead (TEL) process for raising gasoline octane available to German refiners through a licensing arrangement. Interestingly, DuPont, the 50–50 owner of Ethyl, was against giving the Germans the information on TEL manufacture, but it was overruled and convinced by Exxon to go along with this technology transfer. Conversely, the I.G. licensed its *Hydroforming* process for raising gasoline octane through rearrangement of the naphtha molecules to U.S. refiners. Exxon, Chevron and Standard Oil (Indiana) used this technology in time for making high octane aviation gasoline.

War broke out in Europe in 1939, but the U.S. was not yet at war with Germany or Japan. However, it was obviously time for both sides to break up the partnership. This was done in a meeting in The Hague. Both sides kept their information and did not owe anything to the other.

When the U.S. entered the war in 1941, the U.S. government seized the I.G. information relevant to the U.S. war effort and entered into an anti-trust action against Exxon, which was settled soon thereafter. By this time, U.S. refiners had all the information they needed on German hydroforming to produce high octane aviation gasoline. The German patents on BunaS rubber and the knowhow on the use of this rubber for tire manufacture were in U.S. hands. However, the Germans had never passed on to Exxon the knowhow on how to produce the synthetic rubber. This became a challenge for the chemical engineers, as the U.S. rapidly started to build its synthetic rubber industry.

In retrospect, the U.S. gained more than the Germans, since (a) the BunaS technology largely developed in Germany helped the U.S. to build up a massive war machine and (b) German chemical technology on hydrogenation catalysts and in other areas helped greatly in the production of high octane gasoline, as German firms were well ahead in catalysis and high pressure processes. In return. Germany got the TEL technology, information on processing BunaS and the information on making butyl rubber. The U.S. arguably came out the winner in the I.G./Exxon partnership, allowing the U.S. to build a huge synthetic rubber business based on I.G. Farben technology. This story was told in great detail from his company's standpoint in his book entitled *Buna Rubber* by Exxon's Frank Howard, a high level executive of the company deeply involved in the Exxon-I.G. Farben partnerships and technology exchanges. Another book by Joseph Borkin entitled *The Crime and Punishment of I.G. Farben*, showed Exxon in a less favorable light, as far as U.S. interests were concerned, with Exxon more concerned with its business interests than with confronting a Germany preparing for war.

After the war, Exxon came under intense criticism by both President Truman and Congress for cooperating with the Germans. The fact that Exxon's original interest in coal hydrogenation in the late 1920s was in part due to the fact that crude oil supplies were quite short at that time was not considered a good enough reason for spawning a long-term partnership that looked increasingly unpatriotic in retrospect, though immensely important for the war effort.

3.4   A Massive Construction Program Helps Win the War

The U.S. government, notably President Franklin D. Roosevelt, had been warily watching the events in Europe after war broke out in 1939. It was the fall of the Low Countries and France that set in motion an urgent program to prepare the country for war. The Office of Emergency Management was set up in May 1940 to set up and direct emergency agencies to deal with a number of war-related issues, including supply of critical materials. In November the War Resources Board was created under a single administrator. Then on January 7, 1941 the Office of Production Management came into being, created to stimulate production. Among many other things, the office was charged with ensuring the supply of critical materials and on balancing the demands of the military versus the private sectors. This office gave preference to defense orders, covering aluminum, copper, iron, steel, cork, certain chemicals, nickel, rayon, rubber, silk and other materials.

Over the course of the war, new construction activities strongly favored the military sector, as would be expected. Between 1940 and 1942 public construction expenditures rose from $2.6 to 10.7 billion dollars while private construction dropped from

4.2 to 2.7 billion.[15] Residential housing construction for civilians probably came close to a standstill.

The construction of new plants for the supply of various types of war materials, from aircraft, ships, tanks, trucks, explosives, etc., proceeded at a pace never seen before. The government did everything it could to facilitate this effort, making it possible to build plants over periods of less than a year. In March 1941, Edward R. Stettinius Jr., called into high level government service from his position as chairman of U.S.S. Steel Corporation, proposed that plastics could be substituted for aluminum, brass or other strategic materials. With the government offering large new markets, World War II requirements then spurred the production of materials like PVC and polystyrene, which had scarcely been known before. The war effectively transformed these polymers from specialties into commodities.

In 1939, the United States imported about 500,000 long tons of natural rubber. The vulnerability of the nation became apparent, as U.S. industry started to plan the production of synthetic rubber, based primarily on Butadiene-Styrene rubber (BunaS). Exxon, Firestone and U.S. Rubber companies (two licensees of I.G. Farben for the process) had started working by 1940 on plant design and processing techniques. Goodrich was working on BunaN, a butadiene/methyl methacrylate rubber (which it called *Ameripol* rubber) and a rapidly increasing amount of work was being done by companies working on rubber fabricating techniques. Several companies were developing technologies for the production of butadiene, a chemical not even very well-known at that time, but already used in the production of Neoprene. A Polish process for the direct conversion of alcohol into butadiene was evaluated and rejected. It was decided that the two-step alcohol process of Union Carbide was the best technology then available for the production of this key chemical. Research was also accelerated in making butadiene from butylene or normal butane. Styrene was already being produced by Dow, Monsanto and Union Carbide.

A considerable amount of research was done to improve the processing techniques for Buna rubber. A prime discovery was that if polymerization is stopped when only 72% of the monomer is converted and a small amount of mercaptan is added as a modifier, a much better rubber was produced.

In 1941, Exxon was authorized by the Rubber Reserve Program to build a butadiene plant with 15,000 tons capacity based on butylene. Similarly, Union Carbide was to build a butadiene unit with 20,000 tons capacity. Not much later, a number of companies and the government signed an agreement to share all information on production of butadiene from all sources. This was followed by a similar agreement on the production of styrene. In all cases, royalties were set for licensees of either process. It was decided that the new plants were to be designed and erected by private industry but owned by the Defense Plant Corporation, a subsidiary of the Reconstruction Finance Corporation, founded under President Roosevelt's New Deal.

---

[15] Alan G*ropman (1996)* Institute for National Strategic Studies, Washington. *Mobilizing U.S. Industry in World War II. 102.*

When Japan struck at Pearl Harbor, the government immediately called for the construction of plants to make 120,000 tons of Buna rubber by the four companies already engaged on this technology. The government also called for the production of 15,000 tons of Butyl rubber by Exxon. This was somewhat later stepped up to 40,000 tons. In 1942, the president issued an executive order calling for the establishment of a Rubber Director, who would supervise the construction program.

Soon thereafter, the government announced a comprehensive plan for the production of synthetic rubber as follows[16]:

- Neoprene: 40,000 tons per year
- Butyl rubber: 132,000 tons per year
- BunaS: 705,000 tons per year
- Butadiene:

  From alcohol: 242,000 tons per year
  From butane: 66,500 tons per year
  From butylene: 283,000 tons per year
  From refinery off-gases: 113,500 tons per year

Plants were built under the direction of the Defense Plant Corporation, which was under the control of the Rubber Reserve Company. Jim Fair, who worked for Monsanto and was involved in the construction of a styrene plant in Texas City Texas, recalled that they had DPC people on the premises during the construction, but the company worked with Rubber Reserve officials on production matters. The government had purchased an abandoned sugar refinery including a steam plant and used the site and the abandoned office building for the styrene plant Monsanto would build there. The technology came from Dow, whose engineers helped Monsanto to build the plant. Fair described the Friedel Crafts alkylation, which Monsanto had piloted in St. Louis and the ethylbenzene (EB) dehydrogenation unit which involved vaporizing the EB and bubbling it through molten lead (!) Dehydrogenation was at about forty percent. Dow was building a similar unit at Freeport, Texas at the same time. Dow's somewhat unusual dehydrogenation process was described in an AIChE paper in 1945.[17]

By the end of 1944, butadiene was being produced at the rate of 640,000 tons per year and total synthetic rubber from all sources at the rate of 920,000 tons per year! So, within two years, a very large new industry had been created under the pressure of wartime exigencies that could make this happen in a manner not possible under normal peacetime conditions. A map included in Frank Howard's book Buna shows the location of all the plants serving the synthetic rubber program.

Jim Fair described the construction in 1943 of a styrene plant in Texas City, Texas as part of the synthetic rubber program.[18] Dow Chemical had built such a plant before and provided the design as well as giving help in the construction. The ethylbenzene

---

[16]Howard (1947) op.cit. 187.

[17]Fair (1992).

[18]Ibid.

unit was based pilot plant work that Monsanto had done at Dayton Ohio, using a Friedel Crafts type alkylation. Dow supplied the information on the dehydration to styrene. Ethylene was supplied by Union Carbide from their Texas City plant through a pipeline.

The production of what would soon be called "petrochemicals" started ramping up as the United States invested heavily to build synthetic rubber and aviation gasoline plants and in the factories making the raw materials. Between 1940 and 1946, ethylene converted to ethyl benzene rose from 500 to 135,000 tons, ethylene dichloride from 9000 to 27,000 tons, ethyl chloride from 3000 to 28,500 tons and ethylene oxide from 41,500 to 78,000 tons.[19]

The ethylene plants built during the war were mainly sized to serve adjacent derivatives plants "across the fence". However, Union Carbide, which became the largest ethylene manufacturer during the war, built ethylene capacity at Texas City, Texas not only for its derivatives plants, but also for other company's ethylene users. The company was the first to recognize that ethane, the best feedstock for ethylene production, was being separated from natural gas (methane) in order to meet the (lower) heating value required for natural gas serving domestic consumers. So, Union Carbide became a large purchaser of separated ethane, which was much cheaper than propane or naphtha, the other ethylene feedstocks. The company built an ethylene gathering system that collected ethane from gasoline separation plants and supplied the ethane to its ethylene plants.[20]

Toward the end of the war, ethylene had become a major product. Jersey's refinery at Baton Rouge had built a cracker and was the first large scale supplier of ethylene on a merchant basis. By war's end, many cat crackers and naphtha crackers were producing ethylene, propylene, and the other byproducts at a very large rate. Government requirements for these chemicals was dropping rapidly, making them available for civilian uses.

In retrospect, it is easy to see how the government's wartime demand on the oil and chemical industries created the basis and framework for a budding petrochemical industry. The war speeded development of refining and chemical processes, provided feedstocks for petrochemical production and hastened their commercial introduction. A number of companies (Union Carbide, Dow, Exxon, and Monsanto) were building large complexes on the Gulf Coast that were ready to start serving the civilian market. Cooperation and exchange of information had been strongly encouraged by the government to overcome roadblocks to the war effort. Since the government had been sponsoring research, the results of this work were made available to all interested parties through a patent pooling scheme, even though such arrangements were anathema to usual anti-trust arrangements.

The injection of massive amounts of Federal funds to build the refineries, crackers, and synthetic rubber plants redounded to great advantage to the companies that acquired these facilities on very favorable terms after the war, giving companies a

---

[19] *Chemical Economics Handbook SRI International.*

[20] P. H. Spitz op cit. 309.

considerable competitive advantage relative to companies building new plants.[21] The Texas and Louisiana Gulf Coast region became the center of postwar petrochemical production, as the U.S. economy quickly transitioned from wartime to its traditional free market condition.

The creation of a massive synthetic rubber industry, which gave rise to the construction of many crackers and aromatics production units, as well as newly developed polymers and fibers plants in an amazingly short time was remembered by Frank Howard. In his book, he gave much of the credit to the chemical engineers who made it happen. He wrote:

> The American synthetic rubber industry was primarily the creation of the chemical engineer. It is the chemical engineer who has given modern oil and chemical industries the equivalent of the mass production techniques of our mechanical industries. As the name implies, this relatively new profession is the marriage of the pure science of chemistry with the applied sciences of engineering........it is difficult for any race or nation to claim, over a long period pre-eminence in any branch of science. But twenty-five years of intimate contact with the industrial science of Europe left this observer with the conviction that in chemical engineering America has had no close second."[22]

There is no doubt but that chemical engineers were the architects of the petrochemical industry, as suggested in the subtitle of this book. There was great serendipity between the rise and readiness of chemical engineering and the advent of hydrocarbon feedstocks in the 1930s and early 1940s. While chemists invented new molecules for polymers and developed more sophisticated catalysts for chemical reactions, it took the chemical engineering profession and the new ways it had recently defined and adopted to develop new processes for mass manufacture. A brand new industry would be created to produce new consumer-oriented products that transformed the way people lived, travelled and worked.

The development of a massive synthetic rubber and aviation gasoline industry with associated backward integration must be viewed as the first key step in the petrochemical industry's creation. The second step was the commercialization of the many new technologies and the broadening participation of the many new players that entered the industry after the conclusion of the war. It is in that second step that Scientific Design Company led the way.

References

David A. Hounshell, John Kenly Smith, Jr. (1988) *Science and Corporate Strategy* Cambridge University Press, New York. 481.

Edward J. Gornowski (1980) *The History of Chemical Engineering at Exxon.* Advances in Chemistry, Vol. 190. June 1, 1980. American Chemical Society, Washington, DC.

Frank A, Howard (1947) *Buna Rubber*. D. Van Nostrand Company Inc. New York. 13.

Jim Fair (1992) Oral History Program. Chemical Heritage Foundation. Philadelphia, PA 187.

---

[21]Chapman (1991).

[22]Frank Howard op.cit. 241.

Keith Chapman (1991) *The International Chemical* Industry Blackwell, Oxford, UK. 75.
Pap A. Ndiaye (2007) *Nylon and Bombs. DuPont and the March of Modern America.* The Johns Hopkins University Press, Baltimore, MD. 92.
P.H. Spitz (1988) *Petrochemicals. The Rise of an Industry.*. John Wiley & Sons. New York. 70–82.
William F. Furter (1980) *History of Chemical Engineering.* American Chemical Society, Washington, DC. 291.

Bibliography

American Chemical Society. (1977) *A Brief History of Chemistry in the Kanawha Valley.* American Chemical Society, Washington, D.C.
Beaton, Kendall (1957*) Enterprise in Oil. A History of Shell in the United States.* Appleton-Century-Crofts. *New York.*
Dietz, David (1943) *The Goodyear Research Laboratory.* Goodyear. Akron, Ohio.
Ellis, Carleton (1937) *The Chemistry of Petroleum Derivatives.* Reinhold. New York.
Gropman, Alan L. (1996) *Mobilizing U.S. Industry in World War II.* Institute for National Strategic Studies, Washington, D.C.
Morris, Peter J. (1989) *The American Synthetic Rubber Research Program.* University of Pennsylvania Press Philadelphia, PA.
Popple, Charles. *Standard Oil Company (New Jersey) in World War II.* Standard Oil, New York.
Reisch, Mark S. (1998) *From Coal Tar to Crafting a Wealth of Diversity.* C&EN Northeast News Bureau.
Williams, Rever I. ed.(1978) *A History of Technology, The Twentieth Century,* c. 1900–1950. Vol. V. Oxford, Clarendon Press.

Chapter 4
Three Entrepreneurs Join Forces

Abstract Ralph Landau and Harry Rehnberg work together at Oak Ridge, become friends and decide to start a research and development company after the war. Scientific Design Company (SD) is formed in 1946. A third partner, Bob Egbert, also with a Ph.D. in Chemical Engineering from MIT joins SD. A few engineers and chemists are hired and research is started immediately in a small laboratory. Some engineering work is secured to keep the firm going. Information gained from post-war inspection reports on the German chemical industry helps the new firm in accumulating chemical knowhow.

Working together on the atom bomb project at Oak Ridge, Tennessee, two engineers had the idea of starting a company that would carry out independent research and development in the chemical industry. They succeeded, most likely beyond their wildest dreams, and one of the founders, Ralph Landau, became one of the most renowned and wealthiest chemical personalities of his time. This book is a testimony to the success of a very unusual company.

Born in Philadelphia in 1916, Landau attended Overbrook High School in West Philadelphia. He became interested in chemical engineering when he read about this profession, which was being thought of as the new "glamour field". Graduated first in his class, he received a full Mayor's Scholarship to attend the University of Pennsylvania, where he majored in chemical engineering. In 1937, again graduating at the top of his class, he enrolled in the Massachusetts Institute of Technology's graduate school program with a Tau Beta Pi fellowship and he financed much of the rest of his time there as a teaching and research assistant. During a summer internship at the M. W. Kellogg Company in 1939, he worked on the design of petroleum refining plants. There, he started to understand the full potential of the systems approach to chemical engineering".[1] He received an Sc.D. degree in Chemical Engineering from MIT in 1941. While at school, he married Claire Sackler, his childhood sweetheart. They later had a daughter, Laurie.

---

[1]Landau (1994a).

© Springer Nature Switzerland AG 2019
P. H. Spitz, *Primed for Success: The Story of Scientific Design Company*,
https://doi.org/10.1007/978-3-030-12314-7_4

As part of his curriculum, Landau had attended MIT's Chemical Engineering Practice School, where students rotated through several "stations" at industrial plants. An MIT professor at each station posed operating problems provided by the companies as challenges for graduate students to solve over a 6–8 week period. Working in the plants gave these engineers a practical view of how industrial equipment functioned, as well as gaining an understanding of the profit motive in a non-academic environment.

Many of Landau's classmates went to work for oil companies, which were interested in hiring chemical engineers. Some had outstanding careers, including Jerry McAfee, who became chairman of Gulf Oil, Maurice F. "Butch" Granville, who became chairman of Texaco and Robert C. Gunness, who became chairman of Standard Oil of Indiana.

Upon graduation, Landau was hired by M. W. Kellogg as a process development engineer, specializing in the design of chemical plants. A year or so after World War II broke out, Landau was transferred to Kellogg's subsidiary Kellex Corporation to work on the Manhattan Project at Oak Ridge, Tennessee. This transfer was on the strong recommendation by Dr. Warren K. Lewis of MIT, who had been mentoring Landau and believed that he should receive a responsible assignment to help the war effort.[2]

The Manhattan Project, though usually considered as primarily a physics-oriented enterprise, was really also an achievement in chemistry and chemical engineering. In 1940 it was known that an atomic bomb could only work with uranium 235, which must be separated from the predominant isotope, uranium 238 in some manner, posing considerable difficulty. This problem was faced by three extraordinary scientists: Eger V. Murphree (Perkin Medal 1950)[3] president of Standard Oil Development Company, Warren K. Lewis of MIT (Perkin Medal 1936) and George Curme, (Perkin Medal 1935), the engineer who had spearheaded Union Carbide's extraordinary petrochemical research in the 1920s and 1930s. The separation process initially chosen, later supplemented by two other processes, was gaseous diffusion of the uranium hexafluoride isotopes. The government asked Manson Benedict (Perkin Medal 1966), who headed petroleum research at M. W. Kellogg to spearhead the project.

Landau was soon promoted to head up the chemical department at Oak Ridge, where he designed the fluorine plant that was used to make uranium hexafluorides. He also worked on the production of fluorinated compounds such as perfluorocarbons and other compounds, such as plastics and lubricating oils that could resist the fluorine and hexafluorides. In his work on perfluorination, Landau worked with Harold E. Thayer (Chemical Industry Medal 1976), a chemistry professor at Purdue University,

---

[2]Ibid. XVII.

[3]The Perkin Medal is an award given annually by the Society of Chemical Industry (American Section) to a scientist residing in America for an "innovation in applied chemistry resulting in outstanding commercial development". It is considered the highest honor given in the U.S. chemical industry.

who developed purification of uranium oxide, key to the success of the operation. The atomic bomb that fell on Hiroshima contained the radioactive isotope uranium 235.

Harry Rehnberg was born in 1910 in Everett, Washington, the son of Swedish immigrants. His father was an electrician, working on wiring construction projects at some of the many dams in that state. The family was quite poor, scraping out a living. Rehnberg earned a degree in mechanical engineering at the University of Washington, where he also played football. He was known as an extremely extroverted and self-sufficient person. Some people said he was a "hellraiser" and that he didn't like being poor. Upon graduation, he went to work for the Austin Company, which made tanks. When the war broke out, the government asked Austin to provide engineers for the war effort and Rehnberg was tapped to go to work in Tennessee at the Oak Ridge complex.

Working together when building the fluorine plant, Rehnberg and Landau became friends and started to go out to have drinks and they talked about what they would do when the war was over. As they began to know each other, Rehnberg, with his supreme confidence, suggested that the two should go into business together in some way. The two decided that they should form an engineering company to design chemical plants in a manner similar to what Landau had been doing at M. W. Kellogg. Since Landau had worked on organic chemical plants, it would not have been difficult for him to believe that the future of organic chemicals and polymers would shortly be based on petroleum. It was only later that these products would be known as *petrochemicals*. Rehnberg knew absolutely nothing about chemicals, but realized that Landau knew what he was talking about. They quickly saw the logic of becoming involved in the changes that the chemical industry was starting to experience as a result of technologies being commercialized in chemical plants being constructed for the synthetic rubber program and for other wartime needs.

At MIT, Landau had met Robert Egbert who was also studying for an advanced degree in Chemical Engineering. Born in Peru, he had developed an interest in chemical engineering, possibly as a result of his father's influence, who worked as a mining engineer. Both Landau and Egbert had been exposed to the work that Professors Warren K. Lewis and Edward Gilliland were doing on fluid bed catalytic cracking of petroleum fractions, sponsored at the time by Exxon. Gilliland was also known to have had thesis students working on high pressure, very low temperature distillation of low boiling hydrocarbons, which would be useful in designing a fractionation system for future ethylene plants. So, most likely, there had been discussions at MIT about the potential of hydrocarbons as raw materials for producing organic chemicals and plastics. So, when Landau started to think about other people who might be of great help when starting this new company, Egbert came to mind. He had accepted a job at Union Carbide after getting his Ph.D. at MIT and had become well versed in the application of chemical engineering to chemical plant design and operation. Landau thought that Egbert would be an ideal candidate to join the new firm.

Landau was convinced that petrochemicals were slated for very rapid growth after the war. Research based on conversion of petroleum feedstocks had been carried out before and during the war and it tended to be shared across companies, this most likely encouraged by a government bent on winning the war against the Axis through cooperation among scientists. In that way, new "petrochemical" technology came into the Public Domain. Soon after the armistice, Landau became aware of the fact that officers with chemical backgrounds from the U.S., and Great Britain had spent massive amounts of time in Germany after the war, collecting information on German chemical plants. While these plants had largely based their chemicals production on coal, it was clear that some of the German technologies could be used to produce the same products from hydrocarbon feedstocks. This opportunity to obtain chemical process information must have been of great interest to Landau, who most likely applied to obtain copies of relevant reports. With growing confidence and particularly spurred by Rehnberg's optimistic attitude, they felt they had the confidence required to start a company that would enter a new and very promising field.

On June 24th, 1946, Landau and Rehnberg signed a partnership agreement while riding in a taxicab in New York City, a now memorable event recently reported by Jon Rehnberg, Harry's son, who was interviewed for this book. At some point, the company took space at 2 Park Avenue in New York City and rented a small laboratory on 32nd Street nearby. One of the attractions of the laboratory rental was a telephone, apparently difficult to get right after the war.

First, and most importantly, Landau started to talk to Bob Egbert about joining Scientific Design as a full partner. He would bring a great deal of chemical knowledge from his years at Union Carbide, which had much earlier pioneered the production of ethylene as a raw material under the direction of George Curme and had operated a number of other organic chemical plants based on ethylene. Egbert apparently did not need much persuasion and joined as the third partner in September, 1946.

The initial financing of the firm has been detailed in a brief memorandum written in 1988 by Bob Egbert long after his dismissal from the firm. A copy of the memo was provided to the author by Harold Huckins, who kept in contact with Egbert over the years. There were five initial partners, including the three founding executives plus Joseph Skelly and a company called Solvar. That firm owned 10% of the stock and the other four partners each owned an equal share of the balance. Skelly and Solvar presumably put up much of the starting capital to acquire their shares. Within a year, Skelly's shares were reacquired by SD. Solvar's shares were acquired by a Dr. Rosenstein, who later sold his shares to a Dr. Arnold Belchetz. In 1950, Belchetz sold his shares back to Rehnberg, Landau and Egbert. After that, it appears that Rehnberg, Landau and Egbert owned all the shares.

Interested in developing petrochemical technologies and having engineering capabilities related to petrochemical processes, the company looked to hire chemists and engineers who would bring this knowledge to the new firm. The partners saw that the future of organic chemistry and polymers for plastics and fibers production lay in the use of reactive aliphatic hydrocarbons (e.g. ethylene, propylene) and BTX aromatics (i.e. benzene, toluene and xylenes), since these were starting to be used to make products for the war effort in plants that would later make these products for the consumer market. These new raw materials could also be used to make other products with new technologies yet to be developed. That was the direction where the firm was headed, but now how to get there?

The injection of capital allowed SD to start adding staff. One of the first SD hires was Dr. Alfred Saffer, who had been research director at a chemical company after receiving a doctorate in chemistry from New York University. Another was David Brown, who had graduated from Swarthmore with a degree in chemistry and had been a roommate of Egbert's when he was also getting a graduate degree in chemical engineering from MIT. Brown had then joined Shell Development Company at Emeryville, California and had conducted petrochemical research there. He had also been a member of the project evaluation department for processes under development. With this background, he was appointed as the head of SD's process development activity.

Not long after its formation, SD started to contact chemical companies for possible projects. It is highly unlikely that either Landau or Rehnberg had any good connections to potential clients at chemical companies, though the work at Oak Ridge would have exposed them to chemical engineers from such companies, which may have given them some contacts for use after the war was over.

SD's founders had to use their wits to start earning money. A useful approach would be to offer engineering services to chemical firms that had problems in one of their plants. By solving the problem, SD could gain process knowledge, which could then be used to help other companies build other plants. This might be done in partnership with the original company, which in some case might involve sharing the profits with that firm. Or SD might not be subjected to any such a sharing agreement, depending on the details of the contract for the project. The work with Reichhold Company on maleic anhydride, described in Chap. 6 is an example of the latter.

Another way to acquire technology was to benefit from the extensive knowledge that the U.S. gained from Germany, as mentioned earlier. Description of many chemical plants, including operating information, were thus available and sent to Great Britain and the U.S. and they could be accessed and used to design similar plants. Interestingly, Germany had few hydrocarbon resources and so it was no surprise that the Allied officers found that the ethylene, used for styrene production and other syntheses, was produced from coal-based acetylene.[4] In this manner, the Germans were able to effectively make ethylene derivatives from coal!

SD's offer to design and build a phthalic anhydride plant for Witco Chemical in the late 1950s and described in Chap. 6, was almost totally based on details in the report on phthalic anhydride technology used in German plants.

It is therefore likely that Landau and his partners and early hires spent some time in Washington pouring over the so-called B.I.O.S (British Intelligence Objectives Subcommittee) and corresponding C.I.O.S reports prepared by the many British and U.S. chemical executives that were sent to supervise the extraction and copying of the thousands of Bayer, BASF, Hoechst, Chemische Werke Huels, Degussa, Huels, Wacker, and other documents. Much process technology then not in use or not even well known to U.S. chemical companies was there for the taking! And not only flowsheets descriptions and photographs. The C.I.O.S teams were able to obtain and ship back to the U.S. substantial quantities of catalysts. This included, for example, 200 kg

---

[4]Peter Spitz. *Petrochemicals. The rise of an industry* John Wiley and Sons. New York, NY. 5.

of ethylbenzene dehydrogenation catalyst from the styrene works at Huels. Descriptions of various other catalysts used in a number of processes were undoubtedly very useful to the SD researchers.

Of particular interest to Landau might have been B.I.O.S. Final Report No. 1059 entitled "Manufacture of Ethylene Oxide, Ethylene Glycol and Ethanolamines" available from His Majesty's Stationery Office in London for 5 shillings and 6 pence. Amounting to 55 Pages, it gave complete descriptions, flowsheets and economics for the chlorhydrin process, manufacture of glycols and glycol ethers and ethanolamine at I.G. Farben's plant at Gendorf, Germany. In the early 1950s, SD offered to design a plant of this type for ICI in England, no doubt based on some of the information in this report. ICI apparently decided to do it without SD's help. B.I.O.S. Another document, Final Report No. 1052, described the production of acetic acid from acetylene-derived acetaldehyde, as carried out at Hoechst, Huels and Wacker, all members of I.G. Farben. This may also have piqued Landau's interest, since there would be an opportunity to make this important chemical from ethylene instead of acetylene. The reports on German chemistry were therefore a "treasure trove" of information broadly available to chemical researchers (Fig. 4.1).

SD did not originally plan to become a full bore engineering contractor, the type of firm that makes detailed engineering drawings, purchases equipment and builds and erects plants. SD would develop technology at its research laboratory and build a pilot plant to test out new reactions, conduct an economic study to assess the feasibility of the new process, surround the technology with patent applications and look for an industry partner to provide further financing in return for acquiring certain rights. The new plant would then be built by a contractor with knowledge of chemical and refining installations. Old line engineering contractors like Foster-Wheeler,
M. W. Kellogg, Stone & Webster, and Badger were available to be hired by operating companies to build "First-of-a-kind" plants and they were looking for business as the business of building plants for the war effort disappeared. SD would plan to provide a Process Package consisting of a process flow diagram, heat and material balance, and equipment design information. These would then be transformed into a detailed design and equipment purchase list by the contractor selected to build the plant. Since the catalyst employed in the chemical reaction was proprietary, SD would have its catalyst manufactured by a third party, which signed an appropriate secrecy agreement. Eventually, SD became a manufacturer for some of the catalysts required for its processes.

Harold Huckins recently recalled some of the early history of the firm. When Koppers Company, in 1951, was expanding its wartime-constructed ethylbenzene and styrene plant expansion, Huckins, then a Kippers engineer involved in that project, was appointed coordinator for the design of an ethylene oxide plant for Koppers which, however, never was built. Soon after this work, he was asked by Landau to join the young firm, where he reported to Dave Brown, who headed up process development. Familiar with the alkylation technology used by Koppers to convert benzene and ethylene to ethylbenzene, Huckins was able to add this technology to SD's capabilities and knowhow without incurring any restrictions from Koppers.

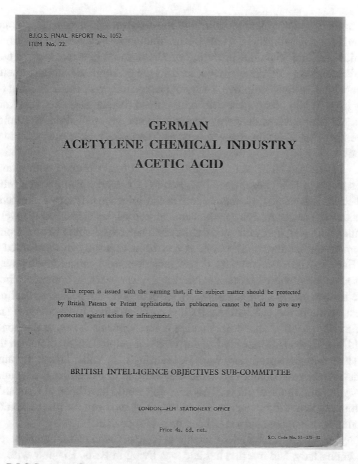

B.I.O.S. FINAL REPORT No. 1052
ITEM No. 22.

GERMAN
ACETYLENE CHEMICAL INDUSTRY
ACETIC ACID

This report is issued with the warning that, if the subject matter should be protected
by British Patents or Patent applications, this publication cannot be held to give any
protection against action for infringement.

BRITISH INTELLIGENCE OBJECTIVES SUB-COMMITTEE

LONDON—H.M STATIONERY OFFICE

Price 4s. 6d. net.

S.O. Code No. 51—275—52

Fig. 4.1 B.I.O.S. report: German acetylene chemistry

According to Huckins, the chief engineer at Koppers had come from a Canadian Government plant in Sarnia, Ontario which had built and operated a synthetic rubber plant during the war. The engineer had been hired by Koppers later in the war to build a similar plant in Pittsburgh. When Huckins left Koppers he was apparently under no obligations not to use his knowledge of the alkylation technology, which was in the Public Domain as far as patents were concerned and where the operating "knowhow" had also been disseminated among the various styrene producers. Thus, SD was able to use this technology for its own projects. Moreover, because of his practical engineering experience, Huckins was asked by Landau to look at the flow-sheet that was being developed on the ethylene oxide process and Huckins made a few suggestions that were immediately accepted.

Researchers and engineers joining SD mostly came from operating companies and brought along with them a substantial amount of "knowhow" covering these companies' technologies and operating experiences. It is a fact that at that time companies tended to be relatively lax in the type of employment agreements they asked new hires to sign. Employees did not change jobs very often, making a career of their jobs. While companies stressed the need for keeping chemical knowhow secret, these admonishments were often not specifically spelled out in employment agreements.[5] That situation changed later, as companies took greater care to protect their knowhow.

When SD started its operations, Europe was in total disarray, with many of its factories partly or totally destroyed by bombing. England's plants were probably in the best condition, but all European countries badly needed to resurrect their chemical manufacturing sector. As European chemical companies started to recover in the aftermath of the war, they realized that they had been left behind in new technology development and commercialization, as had been occurring in the United States. Prewar European chemical research did not include petrochemical development work and this situation obviously continued during the war years, as companies hunkered down to try and survive. So, a number of years of potential petrochemical research were lost, while in the U.S. companies had commercialized technologies and built plants for the war effort that were making petrochemicals. Thus, at the conclusion of World War II, the U.S. already had a network of petrochemical plants on the Gulf Coast while European producers were just starting to consider how to try and catch up to their American counterparts. And they were in dire need for the technologies necessary to build the new plants. This situation would have been in Landau's mind when thinking about likely customers and it may have been for this reason that SD soon found an invaluable person to head up a European activity.

Phil Newman was an American chemical engineer who had fallen in love with a French young lady, had married her and was living in Paris. His new wife was the daughter of a French general with broad contacts in the upper echelons of the French elite, including business leaders. This led to Newman being hired by a leading French firm, Pechiney, soon developing many other contacts in the French chemical industry. Newman was in a sales position at Pechiney and was apparently very successful in that capacity. It is unclear how Newman came to Landau's attention. In any case, Newman joined SD in the late 1940s, in charge of European operations. Newman's contacts with firms like Pechiney, Kuhlmann, Rhône Poulenc and others stood SD in good stead as it started looking for European clients.

Needing funds to support its operation, SD looked for companies that needed help with some of their plants or research activities and this was a way to make the young firm viable, while also developing good contacts. Landau and Rehnberg had little experience in business development, but it turned out that they became excellent salesmen, touting the abilities of their firm and its young and talented staff.

---

[5]Ibid. p. 317.

The partners were also developing the financial model they would use to induce companies to supply the funds needed to prove out laboratory-based work that had the potential of becoming valuable inventions. This would involve successful laboratory research, building and operating a small pilot plant at SD's laboratory, carrying out studies comparing the new technology to current state of the art, and concluding that the process had excellent potential. Millions of dollars would be required to prove out a new process, potentially building a small prototype reactor in an existing plant and then investing in a brand new, "grass roots" plant. SD had little capital, so the money would have to come from development partners. Negotiations would ensue to determine how the company and SD would share in the benefits of successful SD research. In all cases, SD would grant a royalty-free license to build and operate the first-of-a kind plant, and any expansions. What remained to be negotiated was (a) royalty income from licensing other companies, both domestic and foreign, (b) feedback to SD of knowledge gained by the company from the operation of the first plant, including other patentable inventions, and (c) other matters, such as catalyst manufacture. SD's obvious interest was to license as many plants as possible, while the company would want to exploit the new technology without having SD license its competitors.

This was always a complicated negotiation. SD would approach a likely candidate for a partnership in developing the new technology. That firm might have carried out research in the same area, or wanted to be able to carry out such research in the future without being hindered by a non-disclosure agreement that would tie the company's hands. The problem for a process developer, and this was true as much for SD as for any company that was looking for a development partner or just a license, was well explained in an article on Patents and Petrochemical processes by John Kenly Smith.

> Theft of processes is difficult to document, since companies can hide plant technology from competitors. I would also argue that in the negotiating of the sale of a process the buyer should be able to glean enough information to be able to duplicate the process without a license. It appears that the incentive for the buyer to go along with the deal is the prospect of getting the technology in place cheaper than if he did it himself.[6]

Landau quickly became an expert negotiator, working with Tom Gillespie, a tough Irish lawyer whom the company had also hired from Koppers. SD's position was that it could always try to find another financial "sponsor" if a deal could not be made, and this was a real threat to the potential sponsor. SD wanted to make as much money as possible for its invention and could only do that by keeping certain rights for licensing the new technology to other firms for royalties. On the other hand, the sponsoring company was unwilling to give away too many rights to potential competitors, particularly domestic competitors, since it put up most of the money. Eventually, a compromise needed to be and was often found, with SD frequently be discussions with several companies.

The founders spent a lot of time thinking about targets for SD's research. In a sense, the field was quite open, in terms of good avenues for research. An obvious one would be where petrochemical feedstocks would provide better economics than

---

[6]Smith (1998).

existing feedstocks based on coal or alcohol. Replacing acetylene, until then used to make a number of organic chemicals such as vinyl chloride and acetaldehyde, with ethylene might be a fruitful area. Another would be a direct process to replace a multistep process that was then being employed. It would be "breakthrough" research to develop a technology that would provide SD's potential development partner's such dominant manufacturing economics that he could effectively cause existing producers to get out of business, no longer able to compete profitably. This would usually involve a situation where only a few companies were competing, as compared to one where many producers had small market shares. The idea would be to work on technologies where SD could expect to receive a high price for a buyout or substantial royalties, at least from foreign companies (if the partner required a unique domestic position). In such cases, successful research could or would reap rich rewards.

Then, there should be areas where SD might develop technology which, while not "breakthrough" in nature, could still yield engineering income and perhaps some royalties or other fees. This would be a situation where SD could develop technology similar to what was in current use but free from patent infringement. The economics for this case would not necessarily be better than for the current producer(s), but would allow new companies to enter the field. The SD partners also recognized that there were chemical technologies that were more likely to be dominated by engineering contractors whose experience in flowsheet development, equipment design and plant construction would be difficult for SD to emulate, at least at this early stage of SD's business development. Such chemicals as ethylene, ammonia and methanol fell into this category. These chemicals became and remained the domain of engineering contractors, who were selected not so much because of patent ownership, but more because of experience and knowhow.

The three partners fairly quickly decided to start research in two areas. One was the production of terephthalic acid for polyester fiber, currently produced from dimethyl terephthalate in an indirect process based on nitric acid as the oxidant and using methanol to esterify the impure organic acid. SD thought it could develop an intermediate that would use air rather than nitric acid as the oxidant and would produce terephthalic acid directly so as not require use of methanol and its recycle as part of the process. The other target that seemed attractive was the process to make ethylene oxide by direct air oxidation of ethylene. Union Carbide had developed a patented ethylene oxide process and built a plant in the 1930s, using silver catalyst, based on work done much earlier by French and British researchers. If SD could base its work either on expired patents or could work its way around existing patents, great rewards were achievable since Union Carbide was not licensing any potential competitors.

It was time to look for some clients for engineering work that would provide some income and expose SD to the chemical world. Laboratory work was immediately commenced. In 1948 a small pilot plant was running on an ethylene oxide process.[7]

It is clear that the SD management was very gutsy in thinking that it could make breakthrough, patentable inventions that would outpace research done by operating companies with great laboratories and an experienced research staff. But in retrospect,

---

[7]Ralph Landau. op.cit. XIX.

the situation was more promising than it might have looked. Firstly, most companies were initially slow in putting a lot of effort into research based on petrochemical feedstocks, since adopting entirely new technologies would require major changes by companies subject to slow-moving bureaucratic procedures. Secondly, as recounted in several oral histories referred to in this book, companies are often unwilling to spend what is the considerable amount of capital required to prove out a new process that might obsolesce its own technology or does not fit with the chemical areas in which the company is operating. There probably were a number of researchers in different companies who were thinking about novel research along the same lines as the three SD partners, yet did not get support from management to pursue their ideas.

Landau talked about this point in a speech he made to the Newcomen Society in the late 1970s. He quoted an article published in 1978 in MIT's *Technology Review* by Louis Soltanoff as follows:

> If you believe that companies early support the development of new products, you will discover a far different reality. Despite your (inventor's) credentials, you will find that most companies will not jump at your offer to discuss how you could help them meet new product objectives. You disturb them. They prefer their immutable picture of the world in which there is a perpetual demand for their existing product lines, a satisfactory share of the market and a healthy growth in sales and profits. It is a fantasy world in which technical obsolescence never occurs and the competition prices their production fairly. While most companies cannot be described as being innovative, a few are....your problem, of course, will be to locate them.[8]

Landau and Rehnberg surely believed that they could benefit from this corporate mindset at that time.

And then there was another reason why it was easier at this early time in the petrochemical era to bring brand new technology to commercialization. It is much less financially risky to commercialize a new industrial technology when the first plant can be relatively small in size. In that situation, failure is much less of a financial disaster than if the first plant needs to be big enough to compete with large plants using existing technologies.

A related reason why commercialization of new petrochemical technology was easier in the early stages of the industry had to do with **plant size versus time**. The first plants built in the 1950s on brand new technologies were quite small relative to later plants because the demand for petrochemicals was still in an early stage. This favored SD when asking operating company partners to finance the first plant.

A specific example was the first application of SD's nylon intermediates process involving the air-based oxidation of cyclohexane. Monsanto was interested in building a small phenol plant in Melbourne, Australia and here SD saw a chance to commercialize its novel cyclohexane oxidation technology for a different application. This did not turn out well, as described in Chap. 8, but helped SD greatly in terms of proving out the technology of the oxidation process.

---

[8]Landau (1978).

It is important to stress the point about plant size, as alluded to above. Developers of new petrochemical technologies face difficulties in convincing management to risk the much large amounts of money required to build a first-of-a-kind plant for a chemical already being made in a huge plant with the traditional technology. The large plant would enjoy economics of scale that could well compete with a better process if the new process were commercialized in a relatively small plant. The new plant would have a considerable advantage in the cash cost of production (raw materials and utility costs) but the selling price would have to include depreciation and return on investment. This price would have to compete with the cash cost of production of the established producer's partially or wholly depreciated plant. The "bottom line" was that a new process, even when it had production costs that were substantially better than the traditional process, would have to be commercialized in a plant of substantially the same size as operating plants of the day. This fact was and remains an important reason why much less new technology was commercialized as the industry started to reach maturity.

The first few years of SD were very difficult, financially. Claire Landau and David Brown's wife, Jeanine, had small outside jobs, Claire as a social worker, and they helped support their husbands. Landau, in reflecting on the company's early days, said that he did not take any salary for some time, allowing Rehnberg to pay himself a small salary because he already had children. Rehnberg, with his outgoing, self-confident personality was more instrumental than Landau in convincing engineers to leave their jobs and come to work for SD. And when there was no business, Rehnberg was able to convince employees to stay with the firm as "good things were just ahead", even when it was clear that the firm might not survive. Landau also often said that in different aspects of the company's business Rehnberg was always willing to take on the difficult negotiation or confrontation and had no fear (probably a characteristic of his Viking forebears).[9]

Scientific Design Company was unquestionably unique when its founders embarked on an uncertain journey. In later days, Landau said that he and Rehnberg "made up Scientific Design as they went along."[10] They decided right away that they would retain control of the company. They would seek market opportunities and be prepared to take long term risks. Having no capital, they financed part of their early operations with company stock, which they later bought back. They decided that all research would have to have market applications and that research that probably would not be competitive would be quickly abandoned. And they would seek and strongly enforce patent positions.

By 1950 or so, everything was in place and the firm was ready to look for business. Up to the time of the sale of the Mid-Century process to Amoco in the late 1950s, the young company came close to bankruptcy several times, but always persevered. As the saying goes, "the rest is history!"

---

[9]Ibid. 16.
[10]Ralph Landau (1994b).

References

John Kenly Smith. *Patents, Public Policy, and Petrochemical Processes in the Post-World War II Era.* Business and Economic History, Vol. Twenty-Seven, No. 2, Winter 1998. 416

Ralph Landau. (1978) *Halcon International. An Entrepreneurial Company.* Speech to the Newcomen Society. New York 11–12

Ralph Landau. (1994a) *Uncaging Animal Spirits.* The MIT Press. Cambridge, Mass. XVI

Ralph Landau. (1994b) *Uncaging Animal Spirits.*XVIII

Chapter 5
Circumstances Are Perfect: A New Industry Grows Rapidly

Abstract The wartime plants that produced synthetic rubber, butadiene, toluene, styrene, ammonia, etc are auctioned off to private industry by the government. The demand for products made with petrochemicals, as used in housing, automobiles, packaging, vinyl sheets, and textiles, soars and the traditional chemical firms cannot keep up with demand. A number of companies in the rubber, glass, steel, paper, paint and other industries want to enter petrochemicals manufacture, as they see these products substituting for traditional materials. With established producers unwilling to license technologies, SD focusses its research and engineering skills to develop technologies that new entrants can license.

The synthetic rubber construction program that helped the Allies win the war was an immense undertaking. When the war was over, the government no longer wanted to be in a number of war-related businesses, including the chemical business, so it started to look for buyers for government-owned plants that were no longer needed. This was done in the form of auctions. Thirteen styrene-butadiene plants, with a combined capacity of 920,000 long tons had been built by Copolymer Corporation, Firestone, General Tire, Goodrich, Goodyear, U.S. Rubber and three other firms. Two Butyl rubber plants with a combined capacity of 96,000 long tons had been built by Exxon. Nine butadiene plant with a combined capacity of 539,000 long tons had been built by Chevron, Phillips, Exxon, and four other firms. And 194,000 long tons of styrene plants had been built by Dow, Monsanto and Koppers.[1] All of these plants were eventually sold, mostly to the companies that built them, which, in effect, put these companies squarely into the petrochemical business, always at attractive prices. Gordon Cain, a legendary chemical executive and, later, buyer of petrochemical businesses using leveraged financing, recalled the purchase of a government-owned butadiene plant built for the synthetic rubber program. At the time, Cain was working for FMC Corporation . He put in a successful bid after a long negotiation, and bought

---

[1]Petroleum Refiner (1953) Vol. 32. No. 7 July 1953 113.

© Springer Nature Switzerland AG 2019
P. H. Spitz, *Primed for Success: The Story of Scientific Design Company*,
https://doi.org/10.1007/978-3-030-12314-7_5

the plant for eight million dollars! It was subsequently operated as a joint venture between Tenneco and FMC Corporation and called Petrotex Corporation.[2] It was later expanded by installing a Houdry butane dehydrogenation plant when it was found that butane was available cheaply in the summer months. This was a perfect case where a government chemical plant became a mainstay of the new petrochemical industry. Interestingly, another butane-to-butadiene plant was built a few years later when the original alcohol-to-butadiene plants had been shut down. A newly-formed company, Texas Butadiene and Chemical Corporation, was able to obtain process knowhow for such a plant from Chevron, which had built a butane dehydro plant for the synthetic rubber program during the war and was obliged to provide this knowhow to TB&C under government rules.[3]

Monsanto bid and gained possession of the 50,000 metric ton per year styrene plant it had built during the war. The price was $9.9 million versus an original coat of $19.2 million, obviously a bargain, but potential other bidders probably did not try to meet Monsanto's bid. Buying this plant made sense, since Monsanto had started to become a large producer of polystyrene resins. Koppers Corporation also bought the styrene plant it had constructed. DuPont bought a $37 million Neoprene plant for $13.3 million. And Davison bought a $2.1 million cracking catalyst plant for $176,000.[4] It can be assumed that DuPont and other explosives producers bought, at a similarly discounted price, the ammonia and nitration plants used for explosives production when the war was over. And so, as the United States entered peacetime, a large number of plants that had served the military well during the war, were now ready to serve an eager civilian market. A newly termed "Petrochemical Industry" had been born during the war and the new industry had been given an unusually rapid start.

The reasons why petrochemicals achieved such rapid growth in the postwar period has been examined by authors such as Keith Chapman.[5] The main reason was the very large, decades-long unfulfilled demand for consumer goods, such as housing, automobiles, appliances and textiles. First, the Great Depression, with its economic hardships, and then the war, with military priorities, had left the American population eager to replace antiquated or unserviceable items. With the economy booming and with substantial savings squirreled away, consumers were happy to spend their money. Secondly, innovation and substitution provided a plethora of new products now made of petroleum-based feedstocks and previously unknown to consumers. Innovation, in marketing terms, refers to the creation of demand by meeting a previously unsatisfied or unrecognized need. The ability of many synthetic materials both to protect and enhance the appearance of retail goods created a massive expansion in packaging, while the ease with which plastics such as polystyrene could be shaped and formed to produce finely detailed extrusions created tremendous opportunities in the packaging of various types of goods. Replacement of natural materials played an important part, since the synthetic plastics and fibers could be used to make prod-

---

[2] Cain (1992).
[3] Spitz (1988).
[4] Ibid. 154.
[5] Chapman (1991).

Fig. 5.1 Polyethylene bottles and containers. *Source* Getty stock images

ucts that were often substantially less expensive than those made from their natural counterparts. (Fig. 5.1).

Polyethylene bottles and containers replaced glass in many applications at greatly reduced cost. A dramatic comparison showed that the price index for glass containers, paper and paperboard and tin plate rose from 100 to 140 between 1960 and 1970, while for polyethylene and polypropylene the price index dropped from 100 to 50![6] Iron pipe (outdoors) and copper piping (indoors) could in many places be replaced by PVC. Vinyl (PVC) siding was used in housing construction, replacing wood, asphalt or aluminum. Leather upholstery in cars gave way to vinyl. Cellophane film was replaced by low density polyethylene, glass and milk containers by high density polyethylene, wood (e.g. in pallets) by high density polyethylene, and rayon, wool and cotton by nylon and polyester. While in some cases, the new textile materials, though cheaper, were inferior to what they replaced, in others they were better, such as drip-dry and crease-resistant fabrics (Fig. 5.2).

The piping network on the U.S. Gulf Coast gave a great advantage to petrochemicals producers. It made it possible to build large new ethylene plants that could place excess initial production into the merchant market for purchase by non-integrated ethylene derivatives producers. And the piping network, let the interconnected players trade a number of feedstocks and derivatives with each other. All of them benefited greatly from the availability of inexpensive natural gas and gas liquids, a situation that was unique at that time to the U.S. Gulf Coast. The U.S. petrochemical industry was cited by Michael Porter as an excellent example of *national competitive advantage*.[7]

---

[6]Shell International Chemical Company publication (1979).

[7]Porter (1990).

Fig. 5.2 Polyester garments, shoes and handbag. *Source* Getty stock images

The amount of new ethylene capacity built on the Gulf Coast and to a much smaller extent in other parts of the country in the 1950–1970 period was truly astounding. An indication of the enormous growth of the ethylene industry, whose biggest customer was polyethylene, is shown by the rapid ramping up of polyethylene production from 5 million pounds in 1945 to 200 million pounds in 1954 and to 1.2 billion pounds in 1960. Polyvinyl chloride production rose from 1 million pounds to 120 million pounds during the war and to 320 million pounds in 1952. And polystyrene from 15 million pounds in 1945 to 680 million pounds in 1960.[8] Spectacular growth in the use of synthetic fibers is evidenced by the fact that between 1955 and 1975, nylon use increased from 184 to 2288 thousand metric tons while polyester fiber use increased from 16 to 3360 thousand tons, as reported by Stanford Research Institute.

The unprecedented growth in demand for materials produced by the petrochemical industry led to a market growth of about 2.5 times the growth rate of GNP over the 1950–1970 period. This swelling of demand for the bulk intermediate chemicals needed for conversion to petrochemical end products created opportunities for scaling up production processes to levels far greater than anything seen before.[9]

In a book entitled *American Chemical Industry, A History,* William Haynes, who chronicled the U.S. chemical industry wrote, "A new chemical industry was born in the United States between the World Wars. *The commercial production of aliphatic chemicals is the masterpiece of American accomplishment in the chemical realm, a scientific and economic triumph outshining the establishment of the vaunted, better publicized German coal tar chemical industry. The chemical engineering upon which it is based is vastly more intricate, so that its production apparatus must cope with extraordinary pressures and temperatures and exceedingly delicate controls over*

---

[8]Chemical Week (1989) "75th Anniversary Issue" August 2nd.

[9]Rosenberg et al. (1992).

operating conditions. Its output quickly surpassed that from coal tar crudes. Its products serve a far wider area of human needs."[10]

The booming market in products made of petrochemicals quickly caused companies from other industries to find ways to enter the manufacture of one or more of these products. Companies deciding that they must enter the petrochemical industry had a number of rationales beyond the substitution imperative affecting companies in certain industries. Conceptually, there was, of course, a difference between the traditional companies protecting their market domains by changing from traditional feedstocks to hydrocarbon feedstocks and the potential new entrants, who mostly saw good business opportunities. The situation played out differently for all of these firms. Scientific Design Company was in a particularly good position to work with new entrants, whose research and engineering activity was not even oriented toward the petrochemical area. Of course, SD also courted traditional producers when SD was developing a technology that would allow such producers to stay ahead of competitors or even dominate the market. In sum, a great variety of companies decided to enter the petrochemical arena.

An obvious question to ask is: why did DuPont not play a leading role in entering the petrochemical industry in a big way? After all, the company was unquestionably the strongest U.S. chemical company in terms of research and market presence and was already engaged in the production of nylon, polyethylene and various coatings, all of which became important parts of the petrochemical industry. The answer is ambivalence. In 1946, President Walter S. Carpenter Jr. considered petrochemicals "as a problem instead of as an opportunity for DuPont."[11] DuPont had traditionally been interested in transforming cheap raw materials into much more valuable products. DuPont had no interest in backward integration into petroleum products, but was unsettled about the oil companies controlling the raw materials for making petrochemicals. The company foresaw the petrochemical industry as a very likely highly competitive industry as it became commoditized. In this, DuPont's executives had great foresight. Nevertheless, DuPont could not stay out of the game. The company eventually acquired Continental Oil Company to achieve back integration, but not many years thereafter divested the company, realizing that Carpenter had it right after all.

5.1 Why So Many Firms Became Petrochemical Producers

– Oil Companies integrating forward: A number of refiners had seen Exxon and Shell starting to make chemical products and now there was a catch-up situation. Clearly, these companies saw the petrochemical industry as a "value added" situation and they had low value feedstocks (e.g. low octane naphtha, low value refinery propylene) to convert to more valuable products. Building a cracker to

---

[10]Chemical Week op. cit. 24.
[11]Smith (1998).

make ethylene was the most obvious step. Building a naphtha reformer to get into the benzene, toluene and xylenes market was another obvious step. Then, going downstream to polyethylene and other olefin derivatives was often part of the plan.

– Gas Companies integrating forward: Natural gas is primarily methane, but when produced it almost invariably contains ethane and propane and often also butane and higher hydrocarbons called "condensate", recovered as liquids from separation plants built at the wellhead. The heating content of natural gas as produced from the separation plants is higher than methane due to its ethane, etc. content. These so-called gas liquids must be separated to bring the gas to the desired and legislated BTU value of gas going to industrial and consumer uses (home furnaces and appliances). The petrochemical industry serependitiously created a perfect market for these higher aliphatics, and so some gas companies were interested in converting these to ethylene. (Examples: El Paso Pipeline Company, Northern Natural Gas Company, Panhandle Eastern Company).

– Steel Companies looking at the Future: With olefins and aromatics starting to be produced from petroleum feedstocks and steel-making technology changing, with a steadily decreasing need for coke ovens and their byproduct chemicals, steel companies wanted to protect their chemical (from coal) market share and also looked for new opportunities, because the chemical business had higher profit margins. Notably, U.S. Steel Chemicals Corp. was created as a subsidiary by U.S. Steel Company and was given the franchise to build a large petrochemical operation as a diversification for the parent company.

– Consumer-oriented companies integrating backwards: (a) National Distillers and Chemicals Company in joint venture with Panhandle Eastern Pipeline Company. NDCC saw that making ethylene from natural gas liquids was going to force its fermentation alcohol out of the chemical market. The answer: build an ethylene and a synthetic alcohol plant close to the so-called straddle plant that was stripping ethane and higher hydrocarbons from a major transcontinental gas pipeline. This created a petrochemical complex at Tuscola, Illinois, using also licensed polyethylene technology obtained from DuPont. NDCC became a very large petrochemical producer, building plants not only for high pressure polyethylene, but also low pressure, high density PE and later a joint venture in polypropylene with Phillips Chemical. As covered in Chap. 9, NDCC also made a breakthrough invention for the production of vinyl acetate, which it also licensed to Bayer and Celanese. And it became a large producer of acetic acid, using the Monsanto technology. These later plants were built on the Houston ship channel, connected to the grid. (b) Rexall. A company selling packaging materials and paper goods engaged in a joint venture with El Paso Pipeline Company. This led to the construction of a large cracker and adjoining ethylene plant in Odessa Texas, and, later, a plant to produce nylon intermediates, using DuPont technology. The plants were built by Ralph Knight, who had been an executive at DuPont and joined the new firm to create a petrochemical powerhouse. (c) Borden, a company buying large quantities of bottles and other containers and interested in backward integration created Borden Chemical and built a petrochemical complex in Geismar, Louisiana, partly based on acetylene and on hydrogen chloride purchased by tanker from Dow Chemical. (d) Georgia-Pacific decided to protects its wood and paper markets by building a petrochemical complex making vinyl chloride, PVC and synthetic phenol, the latter to support its plywood business. (e) Tennessee Eastman Company (Kodak) created Eastman Chemical and built a petrochemical complex at Longview, Texas making polyethylene for Kodak's film business and a number of organic chemicals. The company already had a history of chemicals and fiber production, based on cellulose acetate. (f) American Can Company entered a joint venture with Getty Oil called Chemplex, located in Iowa, making ethylene and polyethylene, the latter including grades suitable for containers and other packaging items.

- Companies diversifying into petrochemicals: Tire and rubber companies recognized that making petrochemical polymers (principally PVC) could be somewhat synergistic to their traditional business (Goodrich, Goodyear, U.S. Rubber and Firestone). Hercules, an explosives producer, found technologies in Europe and built large businesses in the U.S. (dimethyl terephthalate, phenol, polypropylene).
- Traditional chemical producers protecting their market share or seeing opportunities: These companies had mixed feelings about the advent of the petrochemical era. While they hoped to keep new producers out by being unwilling to license new entrants, they saw the direction the industry was headed so they joined the parade (but much too slowly, failing to keep others out). DuPont, with an existing ICI process polyethylene plant built in wartime, constructed a large ethylene/polyethylene complex in Texas; Monsanto agreed to a petrochemical joint venture with Conoco Oil at Chocolate Bayou, Texas, Allied Chemical built an ethylene and fertilizer complex in Louisiana, Dow built huge petrochemical complexes in Texas and Louisiana.
- Chemical companies entering petrochemicals through original research: One example was Celanese Corporation, which was originally involved with acetate plastics and fibers. The company conducted research on vapor-phase and liquid phase oxidation of propane (later also butane) to make a number of oxygenated chemicals, including acetaldehyde, formaldehyde, methanol and methyl ethyl ketone, as well as higher oxygenates.[12]

The oil companies arguably had the most reasons to diversify their operations, going downstream into chemicals production rather than merely selling hydrocarbons to a growing petrochemical industry. In addition to having control of the raw materials, they recognized that Chemicals and Allied Products (A Commerce Department category) had relatively higher profitability on fixed assets than oil refining.[13] And the petrochemical industry was growing at 10–15% annually. An overview of the different paths selected by several of the oil companies diversifying into petrochemicals included:

- Gulf Oil into the production of cyclohexane, styrene, ammonia, and linear alpha olefins.
- Amoco Chemical into terephthalic acid, high density polyethylene and polypropylene.

---

[12]Peter H. Spitz op. cit 307.
[13]Ibid. 341.

– Phillips Petroleum into high density polyethylene and polypropylene, somewhat synergistic to its synthetic rubber business stemming from the war.
– Shell into ethylene oxide and glycol, phenol, epoxy resins, linear alcohols and alpha olefins and elastomers (Kraton).
– Exxon into low and high density polyethylene, polypropylene, solvents and linear alkyl benzene (for detergents).

All of these companies had ethylene and BTX aromatics plants.

So, how did all these firms obtain the technologies that allowed them to build petrochemical plants? Since internal research had not prepared them to do this and existing producers were generally unwilling to grant licenses, their turned to engineering firms.

Contractors became very successful as licensors for chemicals where patents were less important than design techniques and operating knowhow. In ethylene, for example, M.W. Kellogg, CF Braun, Lummus, and German Linde were fiercely competing to build large crackers and other new plants in a situation where there were few "barriers to entry". These firms did individually create a certain amount of competitive advantage in ethylene production due to flowsheet differences and variations in equipment design (e.g. cracking furnaces, quench boilers for high pressure steam generation). The same was true for contractors competing for construction of ammonia and methanol plants, though MW Kellogg tended to be the leader. All used some ICI technology in synthesis gas preparation.

The opportunity that Scientific Design was significantly able to exploit occurred where companies already producing petrochemicals were unwilling to grant licenses to others, hoping or assuming that potential new entrants could not develop technology that could get around patents and could not acquire flowsheet and equipment information held secret by the existing producers. The latter roadblock turned out to be less of a problem where German plant and operating information became available from the C.I.O.S. or B.I.O.S. reports, though this information still required a considerable amount of engineering work to be needed for developing a process that would get around patents held by current producers. That was a problem that could sometimes be solved through targeted research, for instance if a superior catalyst could be developed that was not covered by existing patents. It appears, however, that companies were reluctant to take this path, with research directors mostly thwarted by conservative management. SD, an entrepreneurial firm not hampered conservative thinking, was not afraid to take risks and so decided to undertake research work that might lead to viable process technologies that could earn substantial income from royalties.

SD evidently realized quickly that it should develop technologies that (a) could be licensed to a large number of domestic and foreign companies regardless of the fact that these firms would compete against each other (ethylene oxide, maleic anhydride, cumene) or (b) technologies that would be unique enough to be of particularly high value (terephthalic acid, nylon intermediates, propylene oxide) thereby limiting licensing opportunities but able to command much higher royalties.

5.2   Europe Goes Petrochemical—A Little Later

Much of Europe's refining and chemical industry suffered heavily in the war and substantial chemical research did not resume until hostilities were over, as discussed in an earlier chapter. This situation had two important consequences, namely that American petrochemical producers established a very strong export position with Europe and that U.S. firms also saw a very attractive market for technology exports.[14]

European firms were at a considerable disadvantage in one respect: Before the establishment of the European Common Market in 1955, manufacturers in a given country largely produced for that country's population size, with some exports. This led to the construction of smaller plants than were built in the U.S. for its large market. As a result, when Europe started building its petrochemical plants, U.S. producers had a distinct economic advantage relative to the smaller European plants, even when including freight and tariffs.

Two European researchers made significant contributions to the global petrochemical industry. Karl Ziegler, working at the Kaiser Wilhelm Institute (later renamed the Max Planck Institute) developed the technology for high density low pressure polyethylene, using lithium hydride to obtain chain growth for the ethylene molecules.[15] The first low pressure, high density polyethylene plant, built by Hoechst, came on stream in 1955. Professor Guilio Natta at Montecatini was able to apply a similar type of reaction to propylene to produce polypropylene. Natta received a patent and, with Ziegler, the Nobel prize for his achievement. Ziegler had not covered the use of propylene in his patent application. Actually, Phillips Chemical eventually showed that it had priority over Natta, but that is beyond the scope of this book.

European process developers made other contributions to petrochemical technology, including:

Phenol from cumene: This process was originally discovered in France, with commercialization in several countries, including France, Great Britain, Canada and the U.S.

Perchloroethylen and Chloromethane: From natural gas and gas liquids: Originally developed in Italy and then commercialized in the U.S. and France.

Nylon 6 from Caprolactam: Originally developed in Germany and practiced in Germany, France and Italy.

Polyurethanes: These condensation polymers of isocyanates and polyesters were first developed in Germany by Bayer.

Mercury cell production of Chlorine and caustic. This type of cell was pioneered in Europe by I.G. Farben companies and Solvay.[16] (It is, of course, not a petrochemical technology).

---

[14]Ibid. 374–375.

[15]Ibid. 332.

[16] *"Europe's Chemical Industry Shows New Vigor" (1957)* Chemical & Engineering News. Feb 25, 1957.

5.3 Great Britain

Similar to the U.S., England needed to develop high octane aviation gasoline for its fighter planes. In 1941, ICI, Shell and Trinidad Leaseholds built a gas oil hydrogenation plant at Heysham which, together with ICI, was able to manufacture half a million tons of aviation fuel per year. After the war, some of the output of this plant was, of course, diverted to petrochemical uses.[17]

While in 1941 ICI believed that petroleum was not of great interest as a feedstock for ethylene, which was then largely produced from alcohol, the company leadership changed its mind during the war. It ceded the production of alcohol-based ethylene to British Distillers Corporation and concluded that its future in organic chemicals would lie in access to crude oil-derived feedstocks.[18]

The company then started discussions with the Anglo-Iranian Oil Company (AIOC) regarding the formation of a joint venture that would produce petrochemicals in Iran. These discussions continued over a couple of years, but eventually foundered. ICI there upon decided to build a light petroleum distillate cracker at Wilton, Yorkshire, with feedstocks supplied by AIOC. The cracker producing 27,000 tons per year of ethylene went on stream in 1953, the bulk going to plants making ethylene oxide, ethyl chloride and ethylene dichloride. AIOC was unhappy about being kept out of the chemical arena which, it knew, was more profitable than selling petroleum products.

AIOC, later called British Petroleum, which was still keen to get into the chemical business and quickly found another partner in British Distillers, founding a joint venture company called British Hydrocarbon Chemicals. A cracker was built at BP's refinery in Grangemouth, Scotland on the Firth of Forth, with plans to make synthetic ethanol and other olefin derivatives. Other companies investing in petrochemicals in England the 1950s included Exxon, Monsanto, Courtauds, Shell and Phillips, the latter building a petrochemical refinery with ICI, in part to supply xylene to ICI's thriving *Terylene* (polyester) business. With a DuPont license for nylon, and Courtauld's acetate and, later acrylic fiber business, England's synthetic fiber industry grew much faster than that of any other European country. Shell started wax cracking at Stanlow to make a detergent. Petrochemicals, Ltd, an independent firm later acquired by Shell, developed the Catarole cracking process, which produced both olefins and aromatics. Apparently an adventurous firm, Petrochemicals was the firm that built a small demonstration plant for SD's ethylene oxide process in or about 1952.

English chemical firms remained wedded to coal as the prime chemical raw material for longer than one might have been imagined, since coal was an important domestic resource. However, as the price of coal rose and miners' strikes became problematical, the availability of cheap Middle East oil became more and more attractive. Then, the use of hydrocarbons rather than coal in organic chemical and polymer manufacture increased rapidly.[19] An important development was the growing installation of catalytic crackers in British and other European refineries, as

---

[17] Reader (1975).

[18] Ibid. 395.

[19] Chapman op. cit. 84.

gasoline demand increased relative to diesel for the automotive market. This effectively increased the quantities of olefins and aromatics available for conversion to petrochemical derivatives. Use of crude oil as a basis for petrochemical production meant that British and other European crackers mostly used low octane naphthas as feedstocks, versus U.S. crackers that were largely using ethane and LPG as a raw material. When North Sea gas became available, British crackers started to similarly use ethane and propane as a feedstock.

5.4 Germany

The German chemical companies were somewhat slower to rush into petrochemicals, in part due to the need for massive reconstruction of their chemical complexes in places like Ludwigshafen, Leverkuesen and in the Ruhr, but also because they were so strongly tied to coal as a chemical feedstock. In addition, the terms of the peace treaty between the Allies and Germany at first prohibited the production of "synthetic oil, rubber, and petrol" so that the reestablishment of a number of chemicals was delayed until these restrictions were relaxed.[20] Interestingly, Chemische Werke Huels, an I.G. Farben subsidiary, which had been producing Buna rubber, in 1945 was already starting to produce plastics and fertilizers, but the other large German firms were slower to switch manufacture to allowable products. However, when the ban on producing BunaS was removed, the large firms were already starting to produce substantial quantities of polystyrene, polyethylene and PVC. Actually, the leadership of BASF, personalities such as Otto Ambros and Felix Ter Meer had, even before the war, been concerned that the German chemical industry had been too dependent on dyestuffs and high pressure chemistry and were counselling that I.G. Farben should become more diversified in consumer products.[21]

Production of synthetics in Germany started to boom in the mid- to late- 1950s. In 1959, production of PVC and copolymers was around 140,000 tons, polyethylene and polypropylene 60,000 tons, and acrylic and methacrylic polymers 25,000 tons.[22] In the same year, German production of all non-cellulosic fibers (largely polyester, nylon and acrylics) was 38,000 tons.

Hydrocarbon raw materials for organic chemicals production in Germany rose rapidly from about 23 to 62% between 1957 and 1963, with a corresponding decrease in production from coal (Fig. 5.3).

In 1960, ammonia and methanol were still made via coal gasification (vs. natural gas in the U.S.) while coal-based acetylene was the primary feedstock for vinyl chloride, acrylonitrile and vinyl acetate. Only about 40% of the total volume of organic chemicals was made from petroleum feedstocks versus 80% in the U.S. at that time.

---

[20]Galambos et al. (2007).

[21]Ibid.

[22]Spitz op. cit. 359.

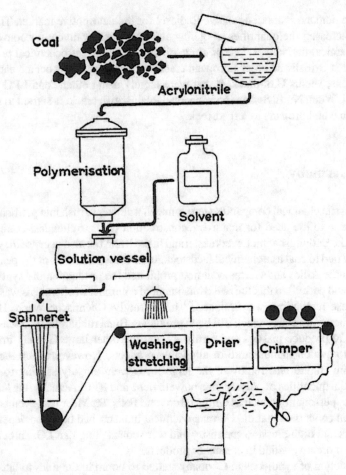

Fig. 5.3 Manufacture of Dralon acrylic fiber by Farbenfabriken Bayer (Spitz op. cit. 285). *Source* John Wiley & Sons

Germany was required to import massive amounts of propylene, butylene, benzene and xylenes, having very little domestic production of these key petrochemical intermediates.

However, the large German chemical firms, such as BASF, Bayer and Hoechst had been reestablished and were in vigorous competition. All of these and smaller firms like Huels, Wacker, Kali-Chemie and Degussa were competing internationally, with some investing strongly in petrochemicals and Germany benefited from a strong economic boom.

5.5 France and Italy

Contrary to the U.S. and Great Britain, the petrochemical industries in France and Italy were established largely as a result of government intervention, tied to helping the economies of less developed parts of the country. In France, the discovery of natural gas at Lacq in Southwest France in the early 1950s led to the formation of Société Nationale des Pétroles d'Acquitaine, which built a cracker complex near the natural gas wells. Several derivatives plants were built there, including production of acetaldehyde, ammonia, methanol and vinyl chloride.[23] Under President De Gaulle, much of the French chemical industry was nationalized, starting with the coal-based chemicals, which were associated with the coal mines, already seized by the government. Chemical executives were recruited by the government to run the nationalized or partly nationalized chemical firms.[24]

The French private companies were relatively slow to invest in petrochemicals, concentrating more on chemical specialties and synthetic fibers. However, a coal chemicals producer H.B.L. at Mazingarbe, which had been a supplier of ethylene, joined forces with Pechiney and Air Liquide to make high pressure polyethylene. Rhone Poulenc built a synthetic phenol plant in 1954. Technology for petrochemicals production came mostly from outside the country, with Kuhlmann and Pechiney licensing processes from Scientific Design. The French government also supported the establishment of crackers in two other less developed regions in the Northeast, namely Carling and the Dunkirk Area. Steam crackers in France were frequently joint ventures, such as the Feyzin cracker near the Belgian border, where the Belgian firm Solvay became a charter member. Exxon built a cracker at the site of one if its refineries at Notre Dame da Gravenchon in Normandy. Petrochemical plastics production in France exploded between 1952 and 1960 from 35,000 to 250,000 tons.

In Italy, somewhat similarly, large amount of public funds were used to build petrochemical complexes in the less developed Mezzogiono regions of Southern Italy, namely at Brindisi near Gela and in Sicily and Sardinia.[25] Anic, a chemical firm that had been established before the war, was acquired by ENI, the government oil monopoly and became Italy's largest producer of petrochemicals.[26] Italian annual steam cracking capacity rose from 16,000 tons in 1954 to 180,000 tons in 1959.

The most prominent Italian chemical company was Montecatini, which expanded its chemical interests and was renamed Montedison when it acquired Edison's chemical interests after Edison was nationalized. It also acquired Snia-Viscosa, an important textile company whose name came from the rayon process. Montedison became an important producer of polyolefins, PVC and other polymers.[27] In 1948, Societa Italiana de Resine (SIR) convinced the government to set up a third petrochemical pole in Sardinia at Porto Torres. Starting with a large phenol plant to

---

[23] Spitz op. cit. 362.
[24] Fred Aftalion (1991).
[25] Aftalion op. cit. 301.
[26] Ibid.
[27] Ibid.

feed its large phenolic plastics business, SIR built a refinery and a cracker, adding polystyrene, low density polyethylene ABS resins, dodecyl benzene, formaldehyde, styrene and acrylonitrile.

With substantial overcapacity looming, the Italian petrochemical industry soon started to head for problems, given the logistical issue of moving product from Sicily and Sardinia to other European countries.

5.6   Japan

The Japanese chemical industry was completely destroyed during World War II. Accordingly, it was decided that rather than developing its own research from scratch, it was better to import ready-made US proven technologies by licensing. The industry was starting to be rebuilt when the Korean War came along, creating a boom in Japan and in its industry.

The country's petrochemical industry was basically orchestrated by Japan's Ministry of International Trade and Industry (M.I.T.I), which asked a twenty member group of high level chemical executives from several Japanese firms, including the Zaibatsu group, to develop a plan. The main conclusion of this committee was that, for expediency, Japan should initially purchase as much petrochemical technology as possible, largely from U.S. firms. The licensing terms were reviewed by M.I.T.I., subject to the approval of the Long Term Credit Bank of Japan whose intention, in part, was for the industry to have balanced locations within Japan. Importing all these petrochemical technologies gave Japanese firms a great deal of knowledge of modern design techniques, equipment and instrumentation. Control by M.I.T.I. gave the licensees a strong negotiating position against U.S. licensors with respect to licensing fees, royalties and terms of contract.[28]

The committee decided on the construction of three petrochemical complexes to be built by Mitsui Petrochemical at Iwakuni on the Inland Sea, by Mitsubishi Petrochemical at Yokaichi, a former navy yard, and by Sumitomo Chemical at Niihama at another location on the Inland Sea. These locations were thus spread out from Hakaido in the north to Kyushu in the south, totaling thirteen units.

From 1950 to 1961, Japanese companies imported 1400 individual technologies and other foreign techniques for a total cost of $300 million. An article by Landau, Tom Brown and Gerson Schaffel, in 1963 described in great detail the evolution of Japan's petrochemical industry, illustrating also the role that SD had in assisting Japan in building its petrochemical industry.[29]

---

[28]Spitz op. cit. 377.
[29]Landau et al. (1963).

Interestingly, the government created incentives for cracker construction by refiners by allowing them to import 3.3 barrels of crude oil for every barrel of naphtha they cracked to petrochemicals. The percentage of imported naphtha use for petrochemical production was almost equal to that used for gasoline. The first Japanese steam cracker, built by Mitsui Petrochemical Company, started up in Iwakuni in 1958, using naphtha as a feedstock.[30] In one case, a U.S. firm (Shell) entered a joint venture with Mitsubishi Chemical to build a petrochemical complex.

Since the U.S. government had not allowed Japan to build large refineries, very little cat cracking capacity had been installed. As a result, there was a limited amount of olefin- and aromatic byproducts from refineries, meaning that petrochemical feedstocks could basically only come from naphtha crackers. Since coke ovens still produced substantial amounts of benzene, Japanese chemical companies at first were suspicious of cracker-produced benzene "since it did not smell right."[31] Initially, there was also an issue regarding propylene. Naphtha crackers' (thermal) cracking operations produce about half as much propylene as ethylene. Yet the crackers are built primarily to make ethylene (for polyethylene, polystyrene and polyvinyl chloride), leaving cracker operators to find a home for the unavoidable byproduct propylene. At the time Japan started building large crackers, polypropylene production was in its infancy, cumene (a phenol precursor and propylene user) was not of great interest in Japan which produced a lot of phenol from coke oven liquids, and propylene oxide was no yet in high demand. All of this meant that cracker-based propylene had fuel or gasoline value, making ethylene economics less attractive than in the U.S., where crackers mostly ran on ethane and LPG, not producing that much propylene. This situation led to Japanese cracker operators desperately looking for petrochemical processes based on propylene.

Over the 1955–1970 period, Japan built a huge petrochemical industry using the most modern technology, which was purchased from the West. The most successful U.S. engineering firms licensing technologies were (a) UOP, which helped Japan rebuild its almost totally destroyed refining industry and (b) Scientific Design Company (a legacy of Tom Brown, SD's senior sales executive, who died prematurely in the early 1970s).

Japanese firms were often able to improve the technologies licensed from the West, making them suitable for re-export. Japanese firms, including the synthetic fiber companies such as Teijin and Toyo Rayon, became very active in developing brand new petrochemical technologies. Examples are acrylonitrile by Asahi Chemical, giant polyvinyl chloride (PVC) plants by Shinetsu, and ABS resins by Japan Synthetic Rubber.[32]

---

[30]Spitz op. cit. 379.

[31]Spitz op. cit. 380.

[32]Saffer and Yoshida (1980).

References

Gordon Cain (1992) Oral History. Chemical Heritage Foundation. Philadelphia, Pa

Keith Chapman (1991) *The International Petrochemical Industry*, Blackwell, Oxford, U.K. 96

Fred Aftalion (1991) *A History of the International Chemical Industry* University of Pennsylvania Press, Philadelphia, PA. 280–281

Galambos, Louis, Takashi Hikino, Vera Zamagni (2007) *The Global Chemical Industry in the Age of the Petrochemical Revolution.* Cambridge University Press, 43

Ralph Landau, T.P. Brown, G.S. Schaffel (1963) *Japan. The never easy struggle toward the top.* Chemical & Engineering News, July 1, 1963. 379

Michael E. Porter (1990) *The Competitive Advantage of Nations* The Free Press, New York, NY

W.J.Reader (1975) *Imperial Chemical Industries* Oxford University Press New York, 364

Nathan Rosenberg, Ralph Landau and David C. Mowery. (1992) *Technology and the Wealth of Nations.* Stanford University Press. 99

A. Saffer and James Yoshida. (1980) *Sources of Technology* Chemtech November. 671

John Kenly Smith (1998) *Patents, Public Policy, and Petrochemical Processes in the Post-World War II Era.* Business and Economic History, Vol. 27, No. 2. 415

Peter H. Spitz (1988) *Petrochemicals. The rise of an industry.* John Wiley & Sons, New York, NY. 314

Bibliography

Barnes, Henry (1975) *From Molasses to the Moon.* The Story of U.S, Industrial Chemical Company.

Brownstein, A.M. (1972) *Petrochemicals. Markets, Technology Economics.*

Hatch, L.F. and S. Matar (1981) *From Hydrocarbons to Petrochemicals.* Gulf Publishing Company, Houston, Tex.

McMillan Frank M. (1979) *The Chain Straighteners.* The Macmillan Press Ltd.

Chapter 6
Scientific Design Company Becomes Successful

Abstract SD is able to obtain some engineering jobs, while the laboratory is developing a direct ethylene oxide process that would be of great interest to new entrants to the industry, as well as to existing global producers currently using the traditional technology. This was SD's fist big success, eventually leading to many domestic and international licenses and engineering projects. SD also develops and licenses several other technologies, including maleic and phthalic anhydride, polyvinyl chloride (PVC), chlorinated solvents, and others. SD is successful in finding new technologies ready for further development in Europe and in granting licenses to European and Japanese chemical clients from offices established in Paris and Tokyo.

Scientific Design Company lost no time in getting started. The first year or two were spent getting organized, taking space for an office at 2 Park Avenue in New York City and renting a small laboratory in a high rise building close by. Then the company started hiring additional staff, partly for research and partly to conduct paid engineering design studies, chemical plant layouts, etc. Some small projects involving engineering work were sold to chemical clients in the first years. There seems no question that the partners must have been somewhat awed by the scope and scale of the adventure they were undertaking. None of them had ever created a business and only Egbert had worked in the chemical industry. They were about to try and prove that a group of engineering and organic chemical entrepreneurs could successfully compete with the research departments of large global chemical companies and also able to convince such companies to form partnerships in the novel technologies SD was starting to think about and create. But if this daunting challenge raised any doubts during perhaps sleepless nights, it did not deter the founders from persevering. Without a broad background in different process chemistries, the partners knew that they had to add to their staff research chemists and engineers who had that exposure and knew the many details that made plants operate. This included equipment design, materials of construction for handling high pressures and temperatures and for corrosive chemicals, commercial equipment for crystallization, filtration and drying, etc. And so they hired people from engineering contractors and operating companies who had designed or worked in plants. And they hired engineers who brought with

© Springer Nature Switzerland AG 2019
P. H. Spitz, *Primed for Success: The Story of Scientific Design Company*,
https://doi.org/10.1007/978-3-030-12314-7_6

them design and operating *knowhow*. Such knowledge would put SD in a position to design relatively similar plants for new clients.

It was immediately evident to the partners that Europe presented a very promising and receptive area for the construction of new chemical plants, given the devastation of the continent during the course of the war and the lack of chemical research performed during the war years. In a speech in 1978, Landau put it this way, "We perceived a need for organic and 'petrochemical' technology as a result of World War II … the greatest area of devastation. (Europe and Japan) offered us broader market opportunities than a more prosperous U.S … and because we knew that innovation and proprietary 'high technology' had been the key to the successful development of roughly comparable companies like UOP and M.W. Kellogg in the petroleum field, (we) started our own original research early in our career.[1]"

Considering the challenges faced by the three entrepreneurs, it is amazing to realize that between 1950 and the early 1960s at least 25 chemical plants designed by SD were built by companies in the U.S., Europe and Australia, about half using original, patented technologies coming out of SD's research. The most important of these was a heavily patented direct oxidation process to make ethylene oxide. The other process that SD was able to develop to a world class standard was for producing maleic anhydride and fumaric acid. SD also developed technology for the production of chlorinated solvents, such as carbon tetrachloride and perchloroethylene, starting with Italian laboratory information and transforming this original research into a licensable process. SD also became a source of polyvinyl chloride (PVC) technology and designed a number of plants using this process. During the 1950s, SD also developed what became its first significant breakthrough process, the direct oxidation of para-xylene to terephthalic acid, the main raw material for polyester fibers, This technology was sold to Amoco Chemical in 1958 for its own worldwide use, though SD retained certain foreign licensing rights and would do the process design for Amoco plants. The so-called Mid-Century process was widely acknowledged as a major coup for SD and put the company on the map as a force in petrochemical innovation. The brilliant work on developing a process that would eventually render obsolete DuPont's and ICI's nitric acid-based route to polyester fiber production is described in Chap. 7. Then, in the 1960–1975 period, SD developed and, with company partners, commercialized three breakthrough technologies: (1) cyclohexane oxidation for polyamide (fiber) monomer, (2) a completely novel direct propylene oxide process, (3) a three-step process to make isoprene from propylene, using original catalytic reaction systems. This was truly remarkable and carried out by an unusually creative team. Added to SD's success in the development and broadscale licensing of ethylene oxide and maleic anhydride technology, it was a chemical *tour de force* arguably never achieved before or after (Fig. 6.1).

---

[1]Landau (1978).

Fig. 6.1 The two original founders. *Source* Scientific Design Brochure (1962)

6.1 Scientific Design Becomes a Developer of Petrochemical Technology

Chloroacetic acid: The first project accepted by the new firm was a commission by the venerable Hans Stauffer, head of eponymous Stauffer Chemical Company. As it turned out, Landau's former boss at Kellogg had become the head of research at Stauffer and was looking for someone to help design and build a chloroacetic acid plant. By coincidence, Landau called on him around that time, wondering whether there might be a project for the new firm. Without any experience with this chemical, Landau convinced his old boss that the new firm could undertake this project. Stauffer Chemical was one of several relatively small chemical plants still under the control of its original founder. It manufactured a number of specialty chemicals, many based on chlorine, which was also produced by Stauffer. Chloroacetic acid is produced by the chlorination of acetic acid in the presence of acetic anhydride, though other production routes have also been used. It has a variety of uses in the production of drugs, dyestuffs, and thickening agents such as carboxymethyl cellulose. When Stauffer became interested in making this product, it was being manufactured by other firms, probably including Monsanto.

The SD-designed plant was built and apparently worked well, but the designers had chosen the wrong construction materials for some of the equipment so that serious corrosion problems developed. Eventually, the plant was shut down and partly scrapped. Hans Stauffer, who probably felt pleased about launching the new company with its first large project, was not happy with the result and never let Landau forget it.[2] Significantly, this unfortunate result for a plant designed by SD was the first and only time this occurred, except for the ethylene glycol plant designed almost thirty years later (See Chap. 12), which suffered a similar result and led to serious financial consequences.

It's worth noting that the knowledge of how to specify the materials of construction for a plant handling chemicals represents an important part of the *knowhow* that is jealously guarded by companies operating a plant. After patent protection, if any, expires, designers of a similar type of plant, who would not have this knowhow, may make serious mistakes in materials selection. The same company had wanted to produce. The same is, of course, also true when a first-of-a-kind plant is designed by the people who carry out the process and engineering design and then select the construction materials for reactors, heat exchangers, piping, pumps, etc. After a plant is successfully placed in operation, a company gains *operating knowhow* that must be carefully guarded. As will be seen in the case of phthalic anhydride discussed later in this chapter, this is an area where SD ran into problems. Reverting to the chloroacetic acid plant, one can hypothesize that SD's engineers believed that they could easily design this plant, since the synthesis is fairly straightforward. The problem was lack of experience with construction materials, involving inadequate knowledge of corrosion problems that could occur when the reactive chemicals came in contact with some of the originally specified equipment. When a number of equipment items in that plant experienced serious corrosion, Stauffer gave up the project and decided to use the rest of the equipment for other purposes. If the company had wanted to produce the originally desired chemical, it would have replaced the corroded equipment with different types of materials which, if successful, would then have provided Stauffer with the knowhow required to design and operate a plant to make choroacetic acid. However, Stauffer apparently never again tried to make this product. So, lessons learned, but at some cost to reputation.

Cumene (isopropyl benzene): One of the first applications of the alkylation technology that Huckins brought to SD was the ability to design a cumene plant for Allied Chemical, which was planning to enter the production of phenol, using the Hercules process (phenol was still mostly obtained from coke oven gases at that time). The alkylation of benzene with propylene makes isopropyl benzene (cumene) with some di- and tri-byproduct that is recycled to the reactor. This reaction was piloted in SD's laboratory to obtain the necessary design data. Huckins' knowledge of the ethylbenzene process was obviously useful in developing the cumene plant design. The catalyst used for the reaction was similar to that used for producing ethylbenzene and was readily available at that time. It appears that this project proceeded without difficulty and the plant started up and worked well. The project for designing the cumene plant was apparently the first successful design undertaking for SD after the

---

[2]Landau (1994).

disastrous experience at Stauffer Chemical. It would encourage the small staff and let them feel that SD had the kind of potential they were told about when they were hired. The plant was built at an Allied Chemical site in New Jersey, which made it possible for many of the SD engineers to visit the site when the plant was being constructed. SD assisted with the startup and early operation as part of the contract.

A few years later, SD also designed a cumene plant for British Hydrocarbon Chemicals Ltd., a joint venture on the Firth of Forth in Scotland between British Distillers, an alcohol producer, and British Petroleum Company. SD also designed a cumene plant for Mitsui Petrochemical Company. It is not known whether any of the cumene projects yielded continuing royalties for SD or just involved a lump sum payment for the technology and design work. Apparently, no other cumene plants were designed by SD. However, SD did design several ethylbenzene/styrene plants in the 1960s, using an improved version of the alkylation technology.

Ethylene Oxide: It is clear why the three partners identified the direct oxidation of ethylene to ethylene oxide as a primary target for research and development. All existing global producers of ethylene oxide except Union Carbide (UCC) were employing the so-called chlorohydrin route, which employed chlorine as the oxidizing agent and left calcium chloride as an undesirable byproduct. Union Carbide had built its first plant in West Virginia before the war and was from that point on the lowest cost producer of ethylene oxide and ethylene glycol in the world. UCC was evidently unwilling to license its technology either in the U.S. or for foreign producers, leaving an opportunity for developing a direct route like UCC's for anyone able to get around UCC's patents. This situation intrigued the SD partners and they set about to develop a competitive process that would clearly find a lot of takers.

UCC had apparently not carried out a lot of research on the oxidation process, but had taken a license from the French developer Lefort. Therefore, the SD researchers might have figured that they had as much chance as UCC researchers had in understanding the variables in the oxidation and then in designing an improved catalyst and reaction system. Bob Egbert had worked on ethylene oxide at UCC and apparently thought that SD could develop a process that could work around the UCC reaction and catalyst patents. He also believed that SD could develop a flowsheet that would not impinge on UCC patents. This approach would primarily involve mathematical studies on the ethylene oxide reaction mechanism and other theoretical and practical matters.

The most obvious place to start was on the catalyst. Research work was commmenced under Al Saffer in 1948 in the little laboratory on 32nd street in New York. Kinetic studies were separately carried out on the mechanisms defining conversion and selectivity when considering a so-called "hot spotted" reaction in a multitube reactor involving a number of variables including ethylene and oxygen concentration, tube diameter and length, catalyst size and shape and reactor temperature. Further, there is the need to keep the composition of the mixture of reactants and products out of the explosive range, which involves the use of nitrogen or methane (for oxygen feed) as a "balance gas". This work was carried out over a number of months by Bob Egbert and Bob Davis, another Sc.D. from MIT, who had recently joined the firm. Egbert knew of a Canadian consultant, Dr. Plewis of Queens College, Kingston, Ontario, who had done research work on ethylene oxide catalysts. Egbert asked him to develop several spherical catalysts, use of which, if they worked, would get around a key Union Carbide patent. When Plewis produced samples, SD took them into the laboratory to try them out. The eventual catalyst, which had a sphere with a solid core and a porous outer surface, was optimized and manufactured by the Norton Company. The catalyst, in fact, did circumvent an important Union Carbide patent. There was also research comparing a fixed bed reactor to a fluid bed reactor and a number of other studies. At some point, it became obvious that the laboratory SD was using was unsuitable, due to a leaking floor and the fact that it had to be shut down at midnight and for other reasons. Huckins was asked to find another laboratory and he found a suitable one in Port Washington, Long Island. Work continued at the larger laboratory, where it was found that the optimum design and operation involves a choice between temperatures, pressure flow rate and catalyst under which conditions heat could be removed very smoothly and the maximum selectivity was attained at a given conversion level. This also kept the conditions inside the reactor outside the explosive limits. This finding resulted from a combination of laboratory work and extensive mathematical and optimization studies. Engineering and economic studies showing that undesirable back-mixing occurs in a fluidized bed, which led the SD researchers to conclude that the fixed bed approach would be superior to a fluidized bed reaction system. More than a year of work eventually resulted in a patentable catalyst and tubular reactor design. (e.g. *Process for the Oxidation of Ethylene*. Ralph Landau, assigned to Chempatents, Inc. New York, NY Patented June 26, 1956 2,752,362). Yields of ethylene oxide in the range of 55–60% of theory on a mol basis were obtained and these were reasonably close to theoretical yields.

The SD engineers also developed an ethylene glycol process suitable for coupling with an ethylene oxide plant. Information from German ethylene oxide and glycol plants were probably helpful with this work (Fig. 6.2).

Following the research work, the entire process was visualized and a flowsheet developed that could be evaluated for commercial application. Cost estimates showed that the capital cost of the plant would not exceed that of a chlorhydrin plant. A comparison of raw materials showed that with ethylene prices at 4.5 cents per pound and chlorine at 3.0 cents per pound, a 40 million pound per year direct oxidation plant would have a $1.2 million per year cost advantage over a similarly sized chlorhydrin plant.[3] A number of patent applications were filed under Egbert's Landau's, Saffer's and several others' names, with substantial foreign filings, incurring considerable cost. Everything was now ready for scale-up.[4]

Petrocarbon, Ltd. a British firm in Manchester, was operating an ethylene oxide plant using the chlorhydrin process, which it had licensed from Carbochimique, a Belgian firm. Landau contacted the firm and told the management that they should consider the direct process and that SD had an expert (Egbert) who would design a

---

[3]Ibid., 9.
[4]Landau (1953).

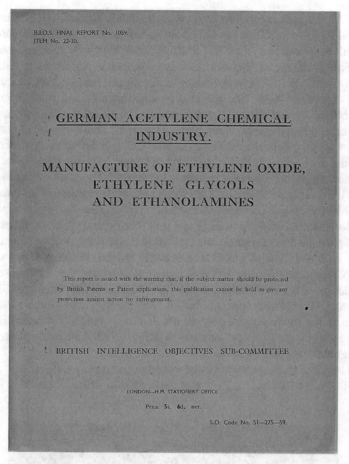

Fig. 6.2 B.I.O.S. report on production of ethylene oxide in pre-was Germany. *Source* British Government Publication

pilot plant for the British firm. Extensive negotiations led to a decision by the firm to build a small demonstration plant for SD's new reaction system in return for an exclusive British license for the technology. This was SD's first "proof of concept", the fact that operating companies would fund necessary scale-up work on a new process in return for certain rights for application of the results to a new plant using the technology! This was a most important step and so Landau and Egbert both sailed for England. Egbert built and operated the pilot plant, which ran into a lot of difficulties, but he eventually successfully demonstrated the reaction system. Egbert then designed a commercial plant, but Petrocarbon ran into financial difficulties and was acquired by Shell. That company decided not to be bound by agreements that Petrocarbon had signed, eventually carrying out its own independent research on ethylene oxide and becoming a competitor to SD.

So SD had to look for another firm for commercialization of its technology, now armed with a considerable amount of operating information from the demonstration plant. Phil Newman, in France, had a strong connection with Péchiney, one of the largest French chemical companies. British Petroleum at that time was building an ethylene cracker at Lavera in the South of France and was looking for customers to build derivative plants. The specifics on how this all came about are not well known, but in any case BP and Pechiney decided on a joint venture, Naphtachimie, to build an ethylene oxide plant using the SD process. This took a lot of convincing by Egbert, dealing with Marcel Pellegrin, who had operated a chlorhydrin ethylene oxide plant for Kuhlmann. Egbert travelled to France and built a large pilot plant at a Pechiney site, which was later transferred to Lavera. The plant was started up by Egbert. A glycol plant had also been designed, possibly using some of the information available in the BIOS report referred to earlier. The ethylene oxide/glycol plant was successfully commercialized in 1953 and was later expanded several times. Egbert also designed the first ethylene oxide catalyst manufacturing plant at Naphtachimie and then designed a similar plant for SD's newly formed subsidiary Catalyst Development Corporation (In later writings, Egbert recalled that the funds SD received for the design of the Petrochemicals and, later, Naphtachimie plants kept SD out of bankruptcy!). What also helped was that, in 1952, the Koppers Company had bought a license on the ethylene oxide process which also helped SD's financial situation. SD was asked by the Koppers Company to design an ethylene oxide plant, but this was never built (Fig. 6.3).

SD's ethylene oxide process was described in an article published several years later). SD granted its first U.S. license of the process to Allied Chemical in Bayport,

Fig. 6.3 Napthachimie's ethylene oxide plant started up in 1953. *Source* INEOS AG

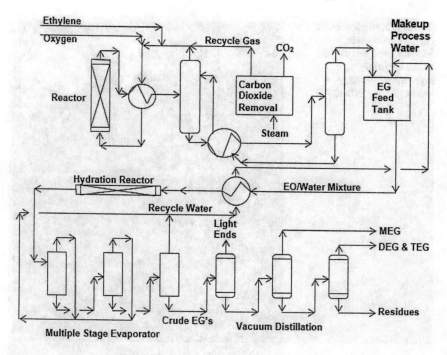

Fig. 6.4 Ethylene oxide and glycol process: simplified flowsheet

Texas and a second one to GAF Corporation (An I.G. Farben successor firm) in New Jersey (Fig. 6.4).

This would have been the time when Union Carbide (UCC) might perhaps have sued SD for patent infringement, but it apparently did not do so or else it did and no information remains about that. UCC had, in 1941, successfully sued U.S. Industrial Chemicals for infringing the LeFort patent for which UCC had acquired an exclusive sublicense, though U.S.I. in its defense had asserted seven grounds in support of its claim of invalidity of that patent.[5] According to Joe Pilaro, a retired U.S.I. executive who was recently asked about this, U.S.I had been marketing an ethanol-based antifreeze liquid and in 1938 changed to a glycol-based antifreeze solution and was therefore doing research on an ethylene oxide process. U.S.I. dropped the project after the patent office action. The likely reason for inaction by UCC relative to SD's ethylene oxide initiative and patent applications may well be that by 1953 the LeFort patent may have expired. In that case, UCC's other still valid patents would have been narrower, such as for catalyst composition or specific reactor design, which would have been easier for SD to circumvent. A second European license was granted by SD

---

[5]Healy (1941).

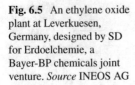

Fig. 6.5 An ethylene oxide plant at Leverkuesen, Germany, designed by SD for Erdoelchemie, a Bayer-BP chemicals joint venture. *Source* INEOS AG

to Societé Chimique des Derivatives de Petrôle (Petrochim) in Antwerp, Belgium. In 1957, SD commenced manufacturing ethylene oxide catalyst in its own factory, rather than contracting to an outside firm. In addition to making the process more profitable for the firm, keeping catalyst manufacturing it in-house was a better way to protect the catalyst knowhow, which soon involved more than one catalyst type. SD also entered the business of regenerating the spent silver catalyst returned by the licensees. It also commenced the practice of leasing the catalyst to its licensees, if desired, thus greatly reducing the capital required companies needed to enter manufacture. SD would finance the leasing with bank loans. Several other European EO licenses were shortly granted (Fig. 6.5).

In 1958, SD held its first meeting of licensees of its ethylene oxide process at the General Aniline plant in Linden, N.J. The idea was to allow licensees (and SD) to share operating information and for SD to present new developments, including new catalysts. A number of technical papers were presented, translated into German and French. (Such licensee meetings were already or soon thereafter also held by ammonia plant operators.) The government took a dim view of these meetings, since they gave participants an opportunity to share pricing and other marketing information, which is illegal under U.S. law, but no action to ban these meetings was even taken, to the author's knowledge (Fig. 6.6).

Shell Chemical successfully developed a competitive process, entered the manufacture of ethylene oxide and also decided to license the technology to any other interested companies. The process used 100% oxygen rather than air, as was the case for UCC and, initially, for SD. Shell teamed up with an engineering contractor, the Lummus Company to compete directly worldwide with Scientific Design. The Shell process used methane as the so-called "balance gas" taking the place of the nitro-

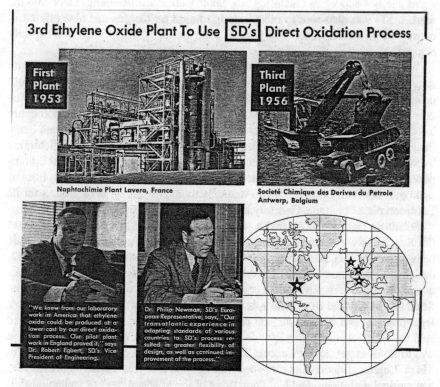

3rd Ethylene Oxide Plant To Use [SD's] Direct Oxidation Process

First Plant 1953

Third Plant 1956

Naphtachimie Plant Lavera, France

Societé Chimique des Derives du Petrole Antwerp, Belgium

"We knew from our laboratory work in America that ethylene oxide could be produced at a lower cost by our direct oxidation process. Our pilot plant work in England proved it," says Dr. Robert Egbert, SD's Vice-President of Engineering.

Dr. Philip Newman, SD's European Representative, says, "Our transatlantic experience in adapting standards of various countries to SD's process resulted in greater flexibility of design, as well as continued improvement of the process."

SD's "International Exchange" Pays Off

So far the payoffs have been in Lavera, France, and the United States. Next year SD crosses the Atlantic again for Societé Chimique des Derives du Petrole's new ethylene oxide and ethylene glycol plants in Antwerp, Belgium.

The less expensive direct oxidation process used in these ethylene oxide plants was developed in SD's American headquarters laboratories and put in pilot plant operation in England.

Naturally, there were many problems in "translating" data—both from

pilot plant to commercial production and from foreign engineering practices to American standards. SD's international experience in ethylene oxide also provided the background necessary to solve these problems of "translation" in chloromethanes, citric acid and maleic anhydride processes.

Today, American companies can take advantage of the continued "international improvement" of SD's lower cost process of direct oxidation of ethylene.

On this or any other organic chemi-

cal plant design problem, SD's services are available to you on a confidential basis. In new plant construction or the revamping of existing facilities to increase production efficiencies, you will profit by utilizing SD's specialized experience in organic chemicals plant design.

SCIENTIFIC DESIGN COMPANY, INC.

Executive Offices:
Two Park Avenue
New York 16, N. Y.
Engineering Offices:
Jersey City, New Jersey

Fig. 6.6 Scientific Design advertisement (ca. 1958). *Source* Chemical Week

gen in an air-based process. This, as previously noted, is necessary for the reaction to take place outside the explosive limit. The Shell catalyst gave somewhat higher selectivities than the SD catalyst, but deactivated a little faster.

The economics of the competing processes were relatively similar. With the Shell process it was necessary to make or buy oxygen, while the SD process required installation of an expensive air compressor and a purge reactor. Lummus and SD competed by offering "packages" that included the engineering design, startup services, catalyst supply and licensing fee. Shortly after Shell/Lummus came into the market, SD was also able to offer a design based on oxygen and also offered to convert its air-based plants to oxygen in some cases, which resulted in a capacity expansion.

With direct oxidation technology economically superior to chlohydrin and with successively larger plants being built and planned, companies still using the old process, such as Jefferson Chemical, decided to withdraw from the business. The only one that did not do so was Dow Chemical, which in Louisiana, had thoroughly integrated its chlorhydrin EO plant with other chlorine-based chemicals and benefited from low cost (to Dow) chlorine. It was able to recycle the ethylene dichloride byproduct of the reaction to its chlorinated solvents unit, where the EDC was decomposed at high temperature to liberate and recycle the chlorine values in the form of hydrogen chloride. However, Dow eventually took a license on SD's process for its European plants in Stade, Germany and in Puertollano Spain.

Scientific Design/Halcon had a very strong legal team and defended its patent position fiercely. An important case in 1966/8 involved the sale of ethylene oxide technology to a Czech company by Snam Progetti, an Italian firm that had built an SD-designed ethylene oxide plant for ANIC, the chemical arm of Italy's government-owned energy giant ENI. This case was described in detail by SD's counsel Stark Ritchie, at that time, who was intimately involved with solving the complex legal problem and obtain financial relief. SD more than suspected that Snam had sold the SD technology in violation of the usual secrecy agreement. However, it was difficult to get good evidence. There were a series of clandestine meetings in Milan, Lugano, London and elsewhere to meet and talk with Italian engineers who had worked with Snam and had knowledge of the alledged theft of technology.[6] It took over two years for the SD lawyers to convince the U.S. Commerce Department lawyers at the Office of Export Control that Halcon had a case and then Halcon sued Snam and Anic for 182 million dollars.[7] Ritchie tried to get help in Washington from a senator and from Clark Clifford, a well-connected D.C, lawyer. Professor E. R. Gilliland of MIT and Dr. Carroll A. Hochwalt, a high level retired Monsanto executive helped as expert witnesses on SD's side. Eventually, the U.S. government officials believed that Halcon had a case and the dispute was scheduled to go to court. While Snam had originally offered one million dollars to settle the case, it was eventually settled for 8.5 million dollars, according to a recent discussion with Barry

---

[6]Richie (1966).

[7] *"Halcon sues ANIC and SNAM for divulging process knowhow"*. Chemical & Engineering News, March 18, 1968. 12.

Evans, an SD lawyer at the time. There were no hard feelings at the end and SD sold another ethylene oxide plant to Montedison, a firm in which ENI had a large interest.

This case served to illustrate Ralph Landau's tough, but essentially not unfair attitude as a business owner. When the Snam case was settled and the money paid, Ritchie believed that he should receive a bonus of close to a million dollars for having worked so diligently over several years and succeeded in winning the settlement. Landau was agreeable to a bonus, but was looking at a much lower figure, telling Ritchie that he was paid a salary to do the job. This dispute eventually went to arbitration, with Ritchie receiving close to the lower number and resigning from the firm.

SD engineers involved with licensing the technology remember a lot of stories connected to their trips. Marshall Frank and several others attended a design conference in Paris with Dow Chemical for an ethylene oxide plant in Europe. Harry Peters, then head of SD's engineering department insisted on this venue because he was adamant not to leave Europe before spending an evening at the Crazy Horse saloon. On another occasion involving an ethylene oxide project in East Germany, Marshall Frank and two other SD people took two dour German engineers to Stenungsund, Sweden on a sales trip, where they could inspect Mo Och Domsjo's SD-designed EO plant. The group subsequently returned to Stockholm and took the visitors to a Folies Bergère type of club. The Germans soon left, muttering about the decadence of Western society. Marshall also recalled a wild ride on slippery roads from Paris to the Northeast region and an EO plant built by Societé Marles Kuhlmann, that was being converted from air to oxygen. The trip was in a small convertible, driven by Maurice Brunet, who had been a manager at Naphtachimie in charge of the first EO startup and had joined SD after that to become part of its EO team. The European office, headed up by Phil Newman, also included Pierre Yakovleff, Andre Dupré, Jean Gignier (ex Pechiney) and Robert Simon, all outstanding engineers who published in several languages.

By 1968, SD had licensed forty EO plants for 21 clients all over the Free World with its process, with the result that its market share for direct oxidation of ethylene oxide processes was 35%, versus 33% for Union Carbide, 17.6% for Shell and 14.3% for all others. According to a brochure published by SD at that time, total annual capacity for SD plants in operation or under design in that year was 2.59 billion pounds.

Maleic Anhydride: Before World War II, this chemical was recovered as a byproduct of phthalic anhydride produced from coke oven-based naphthalene. Maleic anhydride is an interesting chemical, undergoing the so-called Diels-Alder reaction to produce an intermediate for a number of pharmaceuticals and pesticides for which a Nobel prize was awarded to the eponymous inventor in 1950. After the war, unsaturated polyesters came into widespread use, for example for boats and various solid structures and this greatly expanded the market for maleic anhydride. This chemical can be hydrogenated to 1,4-butanediol, used in the production of polyurethanes and polybutylene terephthalate resins.

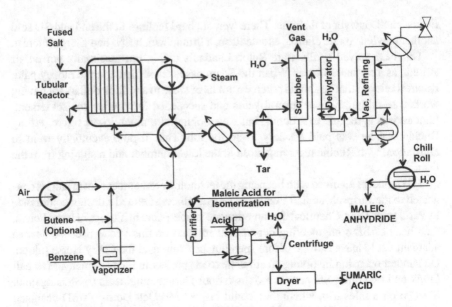

Fig. 6.7 Flowsheet for production of maleic anhydride and fumaric acid

In the mid-1950s, Reichhold Chemical built an "on-purpose" maleic anhydride plant based on benzene, but ran into operating difficulties. Scientific Design was requested to come in to solve problems. Egbert was familiar with maleic anhydride technology from his work at Union Carbide and worked on the Reichhold plant. Over a 12–18 month period, this resulted in an improved operating scheme and a better catalyst. The modified plant was then the only U.S. maleic anhydride plant, with SD, who had made the plant operate satisfactorily, obtaining the right to license the technology to other companies, including in the United States.[8] In addition, SD was able to hire Thomas Brown, a Reichhold sales executive. Together with Landau, Brown subsequently headed up much of SD's sales and licensing activities and soon cultivated a number of important relationships in Japan's chemical industry, with SD granting licenses and providing process designs for various technologies (Fig. 6.7).

The first maleic anhydride license was for a French firm, Compagnie Française des Matières Colorantes, a subsidiary of Kuhlmann, one of the country's largest chemical companies. The design for this plant, which incorporated all the changes made to the original Reichhold plant with added chemical engineering and general industrial knowhow, was carried out almost single-handedly by Bob Egbert. He had been involved in the work on the Reichhold plant, which included a salt-cooled reactor and a distillation system. Before the use of computerized plant design software was in use, he sat down and designed a new maleic anhydride plant on twenty pages of cross-hatched paper, complete with equipment specifications. It was a "tour de force" that few chemical engineers could then or today manage. The CFMC plant started up

[8]Landau and Robert (1962).

Fig. 6.8   Pfizer's fumaric acid plant in Terre Haute, Indiana. *Source* Pfizer Pharmaceutical Company

and ran well. Thereafter, SD also licensed Koppers, Tenneco, American Cyanamid and Monsanto. In all of these cases, SD offered a design package, construction supervision and startup assistance. In return, it received royalties for a specified number of years. Writing in 1978, Landau stated that SD's maleic anhydride process had been licensed to build 40 plants for 25 companies in 15 countries, accounting for approximately 60% of the world's production of this chemical.[9]

SD collaborated with Petrotex Chemical on the design of a plant to make maleic anhydride from n-butane, a process partly developed by Petrotex. However, it appears that this plant was never built. Several companies, including Monsanto and Chem Systems, did develop processes to make maleic anhydride from n-butane. Chem Systems' process was sold to Amoco Chemical which built a plant in Joliet, Illinois. SD looked at the possibility of using a fluidized bed approach for the production of maleic anhydride. This was abandoned, as yields were not satisfactory. One fluid bed maleic anhydride plant was allegedly built by an unidentified firm, but was apparently later shut down.

Fumaric Acid: This chemical, an ingredient of some polyester and alkyd resins and also used as a food acidulant and for medical applications, was traditionally produced from furfural-based succinic acid. The Pfizer Company, in the early 1960s, was interested whether fumaric acid could be directly synthesized. It contacted SD, recognizing that this chemical is an isomer of maleic anhydride. In a relatively short time, SD developed this process, which involves the hydrolysis of maleic anhydride and its isomerization. The first fumaric acid plant was actually constructed for Heyden-Newport. The Pfizer plant with a capacity of 12 million pounds per year was built by SDPlants in 1964 in Terre Haute, Indiana allegedly for $1.7 million (Fig. 6.8).

---

[9]Ralph Landau. Speech to Newcomen Society. 12.

Chlorinated Solvents: In the early 1050s, SD found out that Allied Chemical was interested in getting into the production of chlorinated solvents by direct chlorination of hydrocarbons. Up to this time, these solvents, such as methyl chloride, trichloroethylene and carbon tetrachloride, were manufactured from acetylene, but some scientists thought that they could also be made via high temperature thermal chlorination from methane. Allied asked SD management whether the firm could find such technology. As a result of a concerted search, SD's European office determined that Montecatini had been researching such a process at its Ferrara laboratory, prompted by the fact that natural gas had recently been discovered in the Po Valley. The process needed more work and SD took the information back to its Port Washington laboratory and built a small pilot plant. When the technology was proved out, SD designed a plant for Allied Chemical at Moundsville, West Virginia. It went on stream in 1954 with relatively little difficulty. The new process cut hydrocarbon costs by 75% (!).

In 1955, Péchiney asked SD whether it could design a plant to make perchloroethylene, a dry cleaning fluid, with its chlorinated solvents technology, using liquid hydrocarbons and chlorine. A plant was designed for installation at St. Auban, a small town in the South of France about 100 km north of Nice. After starting up, it ran into severe operating and corrosion problems. Paul Monroe, Vice President of Engineering was sent to try and solve the problems, a situation that actually took a number of months. Eventually, several equipment items and piping were replaced with more exotic materials such as glass, graphite and nickel. Eventually the plant started to work properly and in addition to perchlorethylene, it made a very pure hydrogen chloride byproduct which Péchiney used to make vinyl chloride from acetylene.

In later years, these chlorinated solvents were uniquely produced from aliphatic hydrocarbons (rather than expensive acetylene) using the same type of hot chlorinated process developed by Dow and PPG. But SD was first in this field, benefiting from its strategy of looking for new technology in Europe as well as in the United States. SD designed three more chlorinated solvents plant, one for Kureha Chemical in Japan in the early 1960s, another one for a Mexican company and a third for Montedison in the Po region. Ron Cascone recently recalled working with Irwin Margiloff on a computerized modelling program for designing plants that could make either or both carbon tetrachloride and perchloroethylene. This model predated the more complex modelling system later developed by Aspen Technologies. SD was unsuccessful in obtaining other licensees for its chlorinated solvents technology. Chloromethanes are the main ingredients for producing chlorofluoro refrigerants like Freon, manufactured by DuPont and Allied Chemical among others.

Phthalic Anhydride: In the late 1950s, Mr. Robert I. Wishnick, head of his eponymous chemical company Witco Chemical, was interested in back-integrating from alkyd resins to phthalic anhydride. While Chemiebau von Zieren, a German engineering company, had obtained a license from an I.G. Farben successor company to design and-build a plant for this technology, Wishnick was unwilling to pay royalties to a German firm and approached Landau to see whether SD would design and build a 20 million pound per year phthalic anhydride plant for Witco in the Clearing District on the South side of Chicago. SD found that the phthalic anhydride technology, complete with plant design and catalyst information was described in detail in one of the reports[10] brought back from Germany after the war, and so SD accepted the assignment.

The design prepared by SD was closely based on the flowsheet and equipment list as detailed in the German report, except for the reactor, which, in SD's case, used 2000 or so vertical one and one quarter inch steel tubes. Similar to German technology,[11] a high temperature liquid salt solution was used on the shell side to absorb the reaction heat and piped to a boiler, generating high pressure steam to cool the salt before circulating it back to the reactor. It also used a German-design for the so-called switch condenser to sublimate the phthalic anhydride coming out of the reactor, depositing it on the outside of the condenser tubes so it could be melted off and recovered in a cyclical operation (Fig. 6.9).

Startup problems immediately appeared, first in the form of relatively minor but frequent explosions that occurred in the downstream piping and in the switch condenser, where the reactor outlet mixture deposited phthalic anhydride crystals by sublimation on the tube walls. In addition, the welds at the bottom of the reactor tube sheet started to leak salt into the reactor effluent. Basically, the plant was inoperable! Mr. Wishnick came to Chicago and used some imaginative language, accusing SD of malpractice, etc. But the explosions were a puzzle and there seemed to be no solution, with weeks going by without progress. But there had to be a reason and a way to solve the problem (Fig. 6.10).

The startup group and the Witco operators were completely at a loss. The plant manager, an experienced operator Witco had recruited, finally decided to try and find someone who had worked in a phthalic anhydride plant. He contacted a number of people in the industry until he found someone who knew about explosions problems in these plants and what needed to be done. For a few thousand dollars he informed the frustrated SD engineers and Witco operators that the explosions were due to so-called pyrophoric materials formed from organic solids reacting with metal in the reactor tubes. The problem could be solved by initially passivating the reactor and downstream piping with a phosphoric acid solution and then starting up the plant.

The phthalic experienced gained in Chicago illustrates vividly that *knowhow* sometimes may be just as important as patents when a company wants to design, build and operate a plant using a process that is generally in the public domain, but where some special knowledge is required to make the plant work or work well.

But then another problem surfaced: more unexpected shutdowns. Apparently, after several weeks of operation, small amounts of the hot salt fluid had started through the tube welds, again causing periodic explosions. The reactor was shut down and the tube welds examined. More than two hundred of the welds showed small leaks. It was necessary to construct a diagram on cardboard showing all the tubes as circles and

---

[10]British Intelligence Objectives Subcommittee Report No. 984. Supplement No. 1 Manufacture of Phthalic Anhydride by IG Farbenindustrie.

[11]Spitz (1988).

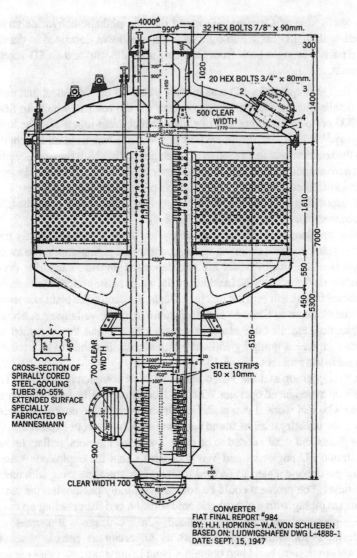

CROSS-SECTION OF
SPIRALLY CORED
STEEL-GOOLING
TUBES 40-55%
EXTENDED SURFACE
SPECIALLY
FABRICATED BY
MANNESMANN

STEEL STRIPS
50 x 10mm.

CLEAR WIDTH 700

CONVERTER
FIAT FINAL REPORT #984
BY: H.H. HOPKINS—W.A. VON SCHLIEBEN
BASED ON: LUDWIGSHAFEN DWG L-4888-1
DATE: SEPT. 15, 1947

Fig. 6.9 Phthalic reactor design used by BASF in pre-war Germany. *Source* U.S. Government FIAT
Final Report No. 984

identifying through inspection where the leaks were occurring. It would eventually be
necessary to reweld all of the tubes in the vertical reactor, using a different welding rod
and flux. It was January, with subzero weather in Chicago—the welders had to stand
on a platform and do overhead grinding and rewelds, a laborious, extremely tiring
and perilous procedure. It would have been easier to take the reactor down and do
the rewelding inside a shop, but the tubes were filled with catalyst and there was no
spare charge and no money to make a second charge of catalyst. So, it was necessary

Fig. 6.10 Witco's phthalic anhydride plant in Chicago, Ill. *Source* Lanxess Aktiengesellschaft

to weld in place. Then, all the welds had to be tested. It took a number of weeks before the reactor was ready and the plant was again stared up. Now both problems were solved—there were no explosions thereafter and the welds held.

This startup was a "baptism of fire" for the SD startup team, as Witco was a firm that mainly made chemicals in small batch kettels, with little or no operator experience operating complex reaction and recovery systems. This was quite different from the startups of ethylene oxide or maleic anhydride plants that were built at companies used to continuous chemical operations. The experience learned solving problems at operating plants at MIT's Chemical Engineering Practice School came in handy for some of the SD engineers starting up the Witco plant. Now, SD had a phthalic anhydride process ready to license to others.

Polyvinyl Chloride and Vinyl Chloride: Vinyl chloride, in the 1960s, became quite inexpensive, as a large number of manufacturers competed on price. A potential PVC resin producer, such as a compounder making PVC piping, could buy vinyl chloride from a number of sources, but needed technology to build a plant. This technology came to SD in the person of Bob Brown. History does not reveal where he had worked before though it is believed to have been one of the rubber companies, most likely Goodrich.

It is very easy to polymerize vinyl chloride, using a suspension process, and the technique was no longer patented by the 1950s. At that time and still today most PVC resin goes into piping, which requires only simple compounding (e.g. coloring, weather resistance) to produce. Compounding tends to be a local business since it is expensive to ship pipe over long distances while vinyl chloride is more easily shipped via rail or truck in large tank cars. Therefore, small PVC plants were often built by compounders in different parts of the country. Some were also built by vinyl chloride producers to move the chemical down the supply chain to compounders who mixed in additives (softeners, colors, etc.) and supplied the finished pipe to specifications to

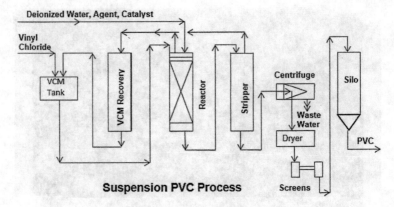

Fig. 6.11 Polyvinyl chloride process—simplified flowsheet

contractors working on municipal or industrial projects. In a couple of cases, piping companies with compounding operations integrated backward into resin production (Fig. 6.11).

SD designed five or more additional PVC plants, including Uniroyal's 50 million pounds per year facility built on a lump sum basis for two million dollars and operated by the Naugatuck Chemical Division. SD also designed a complex for General Tire and Rubber Company at Ashtabula, Ohio, comprising both vinyl chloride and PVC production.

SD was also able to obtain a sublicense from Monsanto for its oxychlorination technology and used it to design a 110 million pound integrated vinyl chloride plant for Mizushima Organic Chemical Company in Japan.

This part of SD's business only lasted for a decade or so. Other PVC technologies were being introduced, either for the production of co-polymers or for mass polymerization (a different technology) and for much larger suspension polymerization reactors, allowing economies of scale. Japanese PVC producers like Shinetsu and Formosa Plastics Company gained a strong foothold in the U.S. by developing these very large reactors. SD, with relatively little polymerization experience eventually decided to withdraw from the PVC arena. While SD conducted original research in many monomer areas, it appears that its scientists never saw an opportunity in developing a new vinyl chloride technology. ICI and others developed a process based on ethane (not ethylene) which was commercialized in the 1980–1990 period. This is an opportunity that SD missed!

Why did firms seek out SD for designing PVC plants? The reason is simple. The other PVC producers were not interested in creating new competitors in a competitive industry. And there were apparently no engineering contractors who could design a PVC plant and provide operating instructions. So, SD filled a gap, consistent with finding opportunities to help additional companies get into the petrochemical industry when existing producers did not want to let them in. In some cases, this involved small companies, such as compounders, located in areas where they could compete as a result of a "freight shield".

The six technologies successfully developed (cumene, ethylene oxide, maleic anhydride, chlorinated solvents, phthalic anhydride and PVC) did not involve fundamental original laboratory research. Different from each other in most respects, these technologies were the result of talented engineers applying their skills to develop licensable processes starting with information available from other sources. In the case of ethylene oxide and maleic anhydride, SD's reaped very substantial rewards through broad-scale domestic and foreign licensing. The technologies described in Chaps. 7 and 8 involved starting with an idea, followed by targeted laboratory research and pilot plant work, and they rank with other breakthrough processes developed by several operating companies during the same period as covered in Chap. 9.

Could other engineering firms have bested Scientific Design by successfully carrying out research in SD's chosen areas. The answer is that they certainly could have. UOP was the firm most like SD, in terms of having a laboratory, process design engineers and a strategy that aimed at generating licensing income. However, the R&D people at UOP were much more comfortable with relatively straight forward processes like alkylation, hydrogenation and crystallization and evidently did not spend a lot of research time on oxidations or complex, multistep processes. The Kelloggs and Badgers were mainly interested in building large plants rather than developing technologies from scratch. There were, however, competitors to SD in terms of looking for third party technologies they could offer. A good example of that is the acrylonitrile process invented by Sohio, (See Chap. 9) where the inventing company chose Badger to build the first and subsequent plants and conferred a sublicense for the technology to Badger so that this firm could build other acrylonitrile plants with that unique fluidized bed process. Badger became known for its expertise in fluid bed reaction system and developed other petrochemical uses for that technique. Kellogg similarly worked with companies including ICI in England to develop large plant ammonia technology and became the go-to contractor for many plants thereafter.

After SD started to become successful, Landau often celebrated the signing of an important contract by inviting a number of his executives and some SD engineers to a dinner at one of New York's finest restaurants. The author recalled dinners at Henri Soulé's Pavillion as well as at Brussels, Chris Cella, Côte Basque and Lutèce. These celebrations had a salutary effect on the organization, which depended on Landau to act as top salesman and keeper of the flame. These elaborate dinners were much appreciated by the staff, given the fact that SD's management was otherwise extremely thrifty, as it had to be, given the lack of available capital. There were periods when there was little work in the shop.

Landau's sales approach in Europe was similarly related to his predilection for great meals at famous restaurants. A world-class epicure, Landau became familiar with many of Europe's three star Michelin restaurants, his favorite being Les Pyramides, near Lyon, where M. Point was a famous chef. At several of these restaurants, Landau entertained the heads of such European chemical firms as Imperial Chemical Industries, BASF, Montecatini, and a number of others. Even if these dinners did not always end up with business deals, Landau helped his other sales executives to receive a welcome reception at these firms. German executives were particularly susceptible to an approach based on superior food and wines. Landau knew that BASF, Bayer and Hoechst maintained famous "cellars" where high level guests were entertained with lavish meals and the best Rheingau wines and this was his way of reciprocating and cementing relationships.

It was important for Landau to develop and maintain relationships with CEO's of U.S. and European chemical firms. He knew that these high level managers were charter members of a very "clubby" environment and Landau was anxious to become a part of the club. This turned out to be difficult for several reasons. Among other things, firms did not put the heads of research and engineering companies on their boards, particularly if such companies were direct competitors with their research establishments. There would certainly also have been some antipathy toward the small, upstart company. What Landau was able to do, however, was to hire some high level retired executives to join SD (or Halcon, which SD morphed into later) to give the company more of an aura of *noblesse oblige*.

SD's success was strongly related to the persuasiveness and risk-taking mentality of its owners and high level executives. Landau promised success and was known to exaggerate the early results obtained in the laboratory, often being proved right by later results. Rehnberg was excellent about closing deals. Tom Brown, was the smooth, persuasive salesman in the classical mold. Dave Brown could charm clients with his good looks and deep knowledge of chemical engineering. Phil Newman spoke perfect French and perhaps other languages and was an experienced head of SD's European operations, with excellent industry contacts. Altogether, it was an impressive team.

The period from 1954 to 1960 saw SD becoming an experienced and innovative engineering firm with a laboratory starting to achieve impressive results. The company was able to gain possession of technologies that could be licensed to a number of companies, giving rise to licensing income as well as engineering jobs. And it achieved its first breakthrough with the Mid Century process described in the next chapter.

References

Gans, Manfred and Brian Ozero 1976. *For EO: Air or Oxygen*. Hydrocarbon Processing, March 1976.

Peter H. Spitz. 1988 *Petrochemicals*. 330.

Ralph Landau. 1953. *Ethylene Oxide by Direct Oxidation*. Petroleum Refiner Sept. Vol 32 146–148.

Ralph Landau. 1978 *Halcon International. An Entrepreneurial Company*. Speech to the Newcomen Society. New York, October. 9.

Ralph Landau 1994 *Uncaging Animal Spirits*. The MIT Press. (11).

Ralph Landau and Robert Simon 1962. Recent Developments in Aromatics Oxidation. Chemistry and Industry Jan. 13, *1962*.

Stark Richie 1966. *History of an Interesting Case*. A personal memoir made available to the author by Barry Evans.

Thomas F. Healy. 1941. *Ethylene Oxide Patent held valid*. Chemical and Engineering News Vol. 19, No. 3.

Chapter 7
The First Major Invention: The Mid Century Process

Abstract SD had, from the beginning, identified polyester fiber raw materials as an area where innovation might be possible. The target was terephthalic acid (TPA), which DuPont and others had not been able to synthesize directly. Creative research and development by SD's chemists and chemical engineers results in the successful, very high yield oxidation of p-xylene to TPA. SD looks for a development partner and contacts Amoco Chemical Company. This eventually leads to Amoco's buyout of this technology, which also had other interesting chemical applications. SD receives a substantial lump sum payment and keeps certain foreign licensing rights. Amoco builds the first plant and then a number of others in the U.S. and abroad, partly in joint ventures, with excellent financial results.

To get started, Scientific Design established itself as an engineering firm, looking for projects where it could sell its process and equipment design knowhow, while also looking for the acquisition of third party technology. However, SD almost immediately rented a small laboratory to start exploratory research. The first goal was work on ethylene oxide production via direct air oxidation and the goal here was to go back to research carried out by others in order to come up with a process that would not infringe existing patents. But the three founders were hoping that SD could also make an exceptional invention and perhaps more than one. Landau, Saffer and Dave Brown undoubtedly spent a great deal of time studying the technology of ethylene, propylene and benzene derivatives to identify areas where original research might lead to a breakthrough invention. The early years of the petrochemical era were a time when a number of existing or new products (mainly polymers) were starting to be produced from hydrocarbons using technologies that were later either improved or optimized or completely changed, as process routes were simplified and made more economical.

And so, SD early on became interested in the technology underlying the polymer that had become the basis of the manufacture of polyester fiber, actually a British discovery. In March 1941, J. R. Whinfield and J. T. Dickson, researchers at the Calico Printers Association had been studying Dr. Wallace Carothers' work at DuPont on condensation polymers which had led to the discovery of nylon, an aliphatic polyester. Whinfield had for some time considered that an aromatic polyester would also have

© Springer Nature Switzerland AG 2019
P. H. Spitz, *Primed for Success: The Story of Scientific Design Company*,
https://doi.org/10.1007/978-3-030-12314-7_7

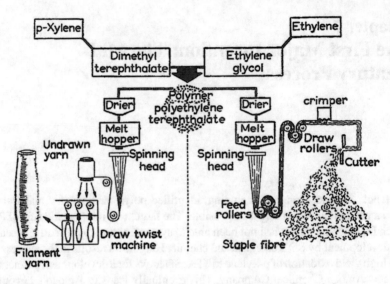

Fig. 7.1 Manufacture of *Terylene* polyester fiber by ICI. *Source* John Wiley & Sons

desirable properties and presented a different opportunity. The two British scientists worked together and soon identified and conducted research on an aromatic polyester that had eluded Carothers, one with a high melting point and other desirable qualities and was based on terephthalic acid (actually dimethyl terephthalate) and ethylene glycol (Fig. 7.1).[1]

The importance of this invention was quickly recognized. Whinfield, at that time was working for the wartime U.K. Ministry of Supply and asked Imperial Chemical Industries (ICI) to evaluate the technology. ICI was already producing nylon in association with Courtaulds and was also a manufacturer of ethylene glycol. This eventually led to ICI buying the rights to the polyester technology from Calico Printers and building the first plant, a 5000 ton per year manufacturing unit making *Terylene*. DuPont had also been working in this area, but found that the British patents obtained by ICI had priority, causing DuPont to pay ICI for a license to make *Dacron*, the DuPont version of polyester fiber. A small DuPont plant making nylon was converted to Dacron in 1948 and the company was now manufacturing both of the new synthetic fibers.[2]

DuPont's manufacturing process involved oxidizing p-xylene in two steps. The first methyl group was oxidized with nitric acid and the second by converting the p-toluic acid formed in this reaction to the methyl ester of terephthalic acid, using methanol. This process was originally the only way to make the polyester, because researchers had found it impossible to oxidize p-xylene directly to terephthalic acid. While the first methyl group was easily oxidized to p-toluic acid, the reaction stopped at that point. Thus, direct oxidation of p-xylene to terephthalic acid was, to that

---

[1] Winfield (1953).
[2] Spitz (1988).

time, deemed not achievable. As a result, all of the plants making polyester fiber used dimethyl terephthalate (DMT) as the ingredient to react with ethylene glycol, requiring recycle of the liberated methanol.

DuPont had actually licensed that two-step reaction from Witten GmbH in Germany, which was later acquired by Dynamit Nobel. That company built a DMT plant and subsequently formed a licensing venture with Hercules, an ICI spinoff in the United States, which contributed to the technology. Still called the Witten process, it was licensed extensively over the next decades in spite of the tremendous success of a direct process to TPA developed by Scientific Design, as discussed in detail in the rest of this chapter.

In 1953, Landau had the opportunity to meet Dr. A. Caress, chairman of ICI and George F. Whitby its managing director and also met Whinfield, who had joined ICI. Undoubtedly, this led to some discussion regarding the indirect process being used to make the aromatic molecule.

Landau and Al Saffer wondered whether a direct route to terephthalic acid might somehow be developed. DuPont had tied up the very limited supply of p-xylene, which was also very expensive, and so the SD researchers scratched their heads to come up with another raw material. Serendipity occurred when the French firm Société Kuhlmann around that time approached SD with a request for developing a process to make terephthalic acid. This energized the new firm to brainstorm alternatives to find a way towards a possible new raw material and technology. Kuhlmann had been looking at alkylation of aromatic substrates with aluminum chloride catalyst to make para-tolualdehyde. SD did accept the assignment, but soon pointed out that the Kuhlmann idea for making terephthalic acid in this manner looked economically unattractive. SD had recently developed a cumene (isopropyl benzene) process and the reseachers thought that this compound could be further alkylated with propylene to made di-isopropyl benzene, actually the meta and para isomers, which could be separated. This, if successful, would then provide a logical feedstock, p-di-isopropyl benzene (PDIPB) which, when oxidized, would produce terephthalic acid. Kuhlmann decided it would financially sponsor such an approach.

Such a route, if successful, would be very attractive economically, at least at then current raw material prices. Benzene and propylene in the early 1950s were selling at 5 and 3 cents/lb. respectively, while p-xylene was priced around 20 cents/lb. This meant that at theoretical yields, terephthalic acid from PDIPB could be made for 3.9 cents/lb raw material cost, versus 12.8 cents/lb. from p-xylene, using technologies in current use.

Laboratory studies were commenced, first in the original small New York City laboratory and then in a new laboratory in Port Washington, Long Island. The first problem was to study the separation of the para- and meta-isomers obtained from the cumene alkylation with propylene. (The ortho-isomer is not formed due to steric hindrance.) It was found that there is only a 6.6 °C difference in boiling point between the para- and meta isomers, but this would be enough for a separation.

Kuhlmann researchers had been following this work and recognized that this approach might work. Initial work looked promising. Kuhlmann then offered to pay SD $50,000 if it could make four successful runs in its pilot plant with Kuhlmann representatives present. This was a time when SD was facing serious financial problems and receiving this payment was critical. The first two runs, under the supervision of Harold Huckins, went well. Then, Egbert showed up, told the others to leave, and decided to personally conduct the third run, which turned out badly. However, the fourth run, supervised by both Landau and Egbert, went well and Kuhlmann decided to pay the money. This experience was the only known case where Bob Egbert's ego got in the way of good judgment.

Kuhlmann then asked for a 300 lb sample of the para isomer so that it could start its own work on converting it to terephthalic acid. SD did not have the equipment to make such a large sample. Huckins, who had recently been appointed to lead terephthalic acid development, contacted Distillation Engineering Company in Caldwell, New Jersey to do the work. A 55-gallon phenolic-coated drum was used as a reactor, with a mixer to alkylate cumene to PDIPB, followed by distilling out the product in the company's distillation equipment. The alkylation procedure took a long time, but cost only $500 (Fig. 7.2).

Over a period of a year or more, the SD scientists then looked at a number of different catalyst systems for the oxidation of PDIPB to terephthalic acid and achieved mixed results. Eventually, Kuhlmann decided that the SD approach was not going to be attractive enough, as they noted that p-xylene prices had started to

Fig. 7.2 Harold Huckins working in SD's office

fall. And so Kuhlmann decided to discontinue sponsorship of SD's work, no doubt to their later regret, when they realized that SD had continued its research and eventually changed its approach on raw material.

SD first continued its research on the PDIPB process, with Bob Barker, a recent hire from Shell Development Company and a key member of the team. Hundreds of experiments using different catalyst systems, solvent media, oxidation initiators and different pressures and temperatures were tried. Disappointingly, it was found that a 60% overall weight yield seemed to be the maximum attainable. This would still be attractive, based on current raw material costs, but it was decided to try still different catalysts and solvents. The researcher had up to that time tried a number of organic salts, but now turned their attention to salts of manganese and cobalt. But now there was an important change in direction. The availability of p-xylene had improved and its price was declining, as more p-xylene crystallization capacity was installed. So, the SD researchers dropped the idea of using PDIPB and switched their research to p-xylene.

At one point, Barker suggested that a bromine salt of manganese and cobalt should be used, but Al Saffer vetoed that ("No bromine in my laboratory!"), according to Huckins, who was working at Port Washington as there was little engineering work going on at 2 Park Avenue. Saffer had gone on a sales trip to Europe and Barker fortunately decided to follow his instincts to try bromine and a manganese catalyst on several model aromatic molecules. It was quickly found that manganese bromide and acetic acid resulted in a rapid oxidation, with practically no intermediate oxidation products in the effluent. Recognizing the efficiency of this system, it was immediately decided to try it directly on p-xylene which, in all previous work by other researchers, had shown to be impossible to convert in one step to terephthalic acid (TPA). On the first experiment with manganese bromide and acetic acid, the reactor outlet system became plugged and the researchers thought that something had gone wrong. It turned out that the experiment had produced a huge amount of TPA. Huckins recalled that when he, Barker and Bob Piesen saw what had happened, they felt that they were witnessing an event like "Edison and the light bulb!" Additional experiments confirmed that this catalyst system gave a 123% overall weight yield (79% of theory) and therefore represented an enormous breakthrough. The researchers concluded that the correct combination of reaction ingredients required for a high efficiency reaction was a source of bromine, a metal catalyst, preferably in the form of manganese or cobalt or mixtures thereof and acetic acid. The reaction was conducted at 400 lb per square inch and 200 °C. These results were quickly confirmed in a pilot plant. Optimization experiments eventually raised the weight yield above 140% (90% of theory).[3]

Engineering studies showed that a commercial plant could be designed with a relatively small reactor and a simplified reaction scheme requiring no recycles.[4]

The next question was how the terephthalic acid from this process would be used to produce polyester fiber. There were enough impurities in the organic acid

---

[3]Landau (1994).
[4]Landau and Alfred (1968).

(e.g. p-toluic acid), to preclude it from being used directly. Producers of polyester polymer at that time used dimethyl terephthalate (DMT) to react with the ethylene glycol, in part because this was the only way to make the polyester, but also because the production of DMT effectively removed most of the impurities, including color bodies. SD's challenge therefore was mainly to match or have less impurities than those in the DMT produced by DuPont and other producers. SD asked the cryogenic laboratory at Pennsylvania State University to carry out a comparison study, using freezing point measurements to assess the impurities in SD-produced DMT against those in commercial DMT. To check this out, the freezing point of then currently manufactured highly pure dimethyl terephthalate was compared to DMT made from terephthalic acid produced in SD's pilot plant. As a result, it was found that SD's DMT was at least comparable in purity to that in current use.[5]

Fortunately for SD, patent applications for the new process, which covered not only terephthalic acid, but other carboxylic acids (e.g. benzoic acid, trimellitic anhydride) had been filed promptly as the work went along. This turned out to be critical, as ICI had recently also been working on a bromine-based oxidation system. Fortunately for SD, it turned out that SD's foreign filings were well ahead of ICI's patent applications.

Now, the question was how to cash in on this spectacular discovery. Landau first contacted the most logical potential partner, DuPont, which was the only producer of polyester fiber at that time. He was met with a firm rebuff. Landau concluded that the reason for lack of interest were two-fold. First, DuPont's Explosives Division was producing the *Dacron* intermediate, using nitric acid for the oxidation and was unwilling to lose this business, which was very profitable for the division. Probably more important was the existence of the famous NIH (not invented here) factor—not surprising when a company of the size and reputation of DuPont has to admit that it was "out-researched" by a small, up-and-coming engineering firm.[6]

Landau then started a series of discussions with ICI which, however, did not lead to an agreement as ICI was doing its own research on a one-step route at that time, realizing that such a route was apparently possible, though SD obviously and fortunately did not disclose any information about its successful research. Eventually, ICI broke off negotiations as its work was starting to get useful results. However, SD's patents had priority over ICI's, resulting in ICI eventually taking a license on the SD technology.

The apparent reason why Amoco became involved is interesting. SD gave a Christmas party for clients and some staff at a fancy New York restaurant. About 100 client executives were invited, including a market development manager for Amoco. At the party, he learned about the terephthalic acid breakthrough and got the word back to Amoco's management.

---

[5]Ralph Landau. op. cit. 24.

[6]Ashish Arora et al. Chemicals and Long Term Economic Growth. John Wiley & Sons. Inc. New York, NY 153.

Amoco Chemical was a large producer of BTX aromatics, and it would have been logical for the company to look at ways to upgrade the xylenes contained in the aromatic BTX mixture. As noted earlier, this was the 1950s and chemical divisions of oil companies were very interested in entering the production of hydrocarbon-based chemicals. At first, Landau and Rehnberg wanted to go the standard route, namely to accept sponsorship of scale-up and commercialization in return for an exclusive license for domestic production. But SD was short of work and very short of cash so when Amoco offered to buy the technology outright, the partners agreed and a price was set. SD was, however, able to keep certain foreign licensing rights.

The buyout price negotiated between the firms was not announced or even known by many people. SD was a private firm and the price was not high enough to warrant including in Amoco financial statements. Rumors had it as "several million dollars", a very low figure, if true, considering that the technology was eventually responsible for a large percentage of Amoco Chemical's operating profits for a number of years. On the other hand, the payout made SD a viable entity, well known to the world's chemical community.

The purchase of the Mid-Century technology by Amoco was a signal event in the then short history of Scientific Design. While a certain amount of engineering work, including for ethylene oxide/glycol and maleic anhydride licenses had kept the young company afloat, the influx of what then was a significant amount of money to SD was a signal event for a company with already a relatively large payroll. It solidified the feeling that SD was here to stay and *primed for success*.

Amoco decided to build a plant based on the purchased technology at its refinery in Joliet, Illinois, which started up in 1958. Against the advice of SD, the plant was designed to oxidize a mixture of xylenes to produce not only TPA but also isophthalic acid (IPA) and phthalic anhydride. The reasons Amoco decided to build a mixed acids plant—far more complicated than a plant for terephthalic acid alone would have been—are unclear. But this approach had been studied in Amoco's laboratory and seemed like a good idea, because the plant would use an inexpensive feedstock (mixed xylenes) readily available from its BTX plant and make Amoco Chemical a supplier of additional products that went into other markets, such as coatings. And the timing would be good because the amounts of naphthalene available for phthalic anhydride production were shrinking, in part due to a prolonged steel strike. So, Amoco went ahead against SD's advice.

The idea turned out badly. This was the most complicated plant that SD had designed to date. The most important problem was the need to separate crystals of TPA and IPA and this was to be done using cyclones. Huckins, who managed the SD team for the startup of the plant, recalled that for weeks the city of Joliet was covered with TPA and IPA dust that emanated from the cyclones. It turned out impossible to make pure products of any of the three acids. Amoco eventually decided to discontinue feeding mixed xylenes and converted the plant to make only TPA (Fig. 7.3).

The first p-xylene oxidation plant by the Mid Century process was actually built by Mitsui Petrochemical Company at Iwakuni, Japan in 1958, with an SD license. This was the first plant to use what later would be called the Mid-Century Process,

Fig. 7.3 Terephthalic acid plant designed by SD for Mitsui Petrochemical Company. *Source* Mitsui Chemical

built before successful commercialization by Amoco Chemical. By the late 1960s, Amoco had authorized an expansion of the plant, while SD had granted several foreign licenses on the technology, with plants in operation or under construction by Montedison, Imperial Chemical Industries, Maruzen Oil Company, Mitsubishi Chemical Company, Kurashiki Rayon and Algemeinde Kunstzijde Unie N. V.[7]

The fact that SD was successful in selling licenses to Japanese firms as early as the late 1950s was a credit to Tom Brown. His Reichhold contacts were probably with Japanese trading firms like those from Mitsui, Sumitomo and Mitsubishi, which were very powerful and imported large amounts of various U.S. petrochemicals to Japan. A small U.S. firm, Fallek Chemical, was exporting BTX chemicals, including p-xylene to Japan and provided additional contacts. Through the trading firms, Brown developed a number of solid relationships with high level Japanese chemical executives, often depending on Shigeru Ishihara, a prominent trader at Mitsui Trading Company, to reach the right parties. This contact led to discussions between SD and Mitsui Petrochemical and resulted in Mitsui taking a license for the Mid-Century

---

[7]Ralph Landau. op. cit. 24.

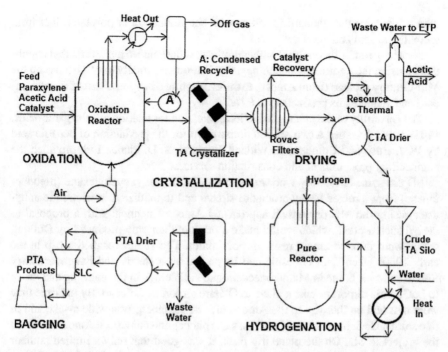

Fig. 7.4 Production of pure terephthalic acid (TPA) via SD's Mid-Century process and Amoco's purification technology

terephthalic acid technology, the first company in Japan to build a plant for this process. Mitsui converted the crude acid to dimethyl terephthalate which was sold to one or more of Japan's fiber producing companies (e.g. Teijin, Toray) for reaction with ethylene glycol to produce polyester fiber (Fig. 7.4).

As noted earlier, the terephthalic acid made by the Mid-Century process was not pure enough to be used directly for making fibers and had to be converted to dimethyl terephthalate to create a monomer pure enough to be reacted with ethylene glycol. Amoco Chemical worked on ways to purify the acid, whose main impurity was p-tolualdehyde. A number of different catalyst systems were tried. There was a thought that treatment with hydrogen would deal with the impurity, but that such an operation would also hydrogenate the benzene ring, so other methods kept being tried. Fortunately, Dr. Herbert Meyer, who had initially worked on the esterification with methanol, found that a palladium-on-carbon catalyst worked perfectly with hydrogen, making a terephthalic acid that was pure enough for direct esterification with ethylene glycol to make the polyester fiber.[8] Meyer became famous for his work, which received a dominant patent. A few years later, he earned the Perkin Medal for his discovery. The product, called PTA (purified terephthalic acid), from

---

[8]Meyer (1997).

then on, became the standard for all new global production of polyester fiber from terephthalic acid and ethylene glycol.

For a long time, other polyester fiber producers continued to make dimethyl terephthalate using the traditional technology. But DuPont eventually took a license on the Mid-Century process (from Amoco, to the everlasting disappointment of Ralph Landau) and switched its production to PTA.

The versatility of the process led to its application for making other organic acids. In 1965, SD designed a two million dollar plant for the production of benzoic acid by W. J. Bush and Company in Widnes, England. S. D. Plants, Ltd. provided the engineering, procurement and construction services.

SD did some laboratory work to recover 1,2,4 trimethyl cyclohexane (pseudocumene) from a mixed heavy aromatics stream and to oxidize it to trimellitic anhydride, using the MC process. It approached Amoco Chemical with a proposal to design such a plant, which would make a product then only produced by DuPont, which went into extremely resistant polyimides. The author recalled a trip in the early 1960s when Tom Gillespie and he picked up a sleepy and hungover Harry Rehnberg at his home in Mamaroneck around six o'clock in the morning and drove to LaGuardia airport to take a plane to O'Hare airport in Chicago. By the time they were in a taxi on the way to the Amoco offices, Rehnberg was wide awake and in fine spirits. He delivered one of his best sales pitches and convinced Amoco to award the project to SD. On the plane trip back, it was good that only a limited number of little martini bottles could be purchased to celebrate the sale. Rehnbert was then delivered back to his home safe and sound. Some work was allegedly also carried out on the application of the Mid Century process on durene (1, 2, 4, 5 tetramethyl cyclohexane) to make pyromellitic dianhydride.

Amoco Chemical built a number of large TPA plants in the U.S., Europe and China, in some cases with partners. Although Amoco was not obligated to use SD for the design of future plants, the company continued to have SD provide the so-called process packages (i.e. the process design, flowsheets and equipment specifications) used by contractors to do the detailed engineering design and procurement for these plants—a source of continuing income for SD.

While Amoco Chemical also manufactured other petrochemicals, such as high density polyethylene, polypropylene and a wide range of aromatics, TPA was the most important chemical for the company, being a differentiated technology with associated high profit margins. Amoco Corporation was eventually acquired by British Petroleum Company (BP) and Amoco Chemical was merged with BP Chemicals which had acquired British Hydrocarbon Chemicals in Scotland and also became a producer of acetic acid and vinyl acetate.

Amoco's purchase of the Mid-Century process from a small research and engineering firm was a coup for Amoco as well. Both Amoco Chemical and later BP Chemical (which acquired Amoco) acknowledged that the TPA technology was the most profitable part of their business for a number of decades, with many owned plants and licensees existing all over the world. In an interesting sidelight, Amoco let it be known that while the Mid-Century process threw off a lot of cash, the company kept plowing money into the construction of plants in the U.S. and abroad, so that it took over ten years before the process was "cash positive".

An important development associated with the sale of the Mid-Century process was the forced departure of Bob Egbert. He had been active in the development of the process, but there had been increasing tension between the three partners. Egbert, who had possibly never been to Europe before joining SD, started to enjoy the life there on his many trips, though he was always doing the sales or supporting work there (Personal recollections, H. Huckins and the author), even when he was enjoying himself in the off hours. Evidently, his life style was anathema to Landau. Marvin Ellen, who was an accountant for SD, recently recalled that around that time Landau decreed that SD executives travelling to Europe or Japan should take their wives along!.

To sum up, Egbert was a very independent spirit and was spending money a little too freely on his many business trips to Europe. The two other partners had become increasingly upset by his personality and behavior and, now with some money in the bank, decided to buy him out and evidently gave him an ultimatum. In a personal written recollection, copy of which was provided to the author by Huckins, Egbert claimed that "he was forced out of the company in 1957 and his shares were purchased by SD for $100,000, net after taxes and legal fees".

The departure of Bob Egbert was a big negative for SD. Egbert was, by many, considered the top process engineer in the firm. His contribution to the Mid Century process was important (three patents), and his role in helping SD to develop the ethylene oxide process and become a worldwide licensor of the ethylene oxide process was crucial to the early survival of the firm and a source of sustaining income. He also claimed to have been the engineer who solved the catalyst problem at Reichhold's maleic anhydride plant, and improved its maleic anhydride purification system, leading to the construction of the first SD maleic anhydride plant in Franc and a strong licensing position by SD thereafter. U.S. patent 2,777,860 issued to R. Egbert and M. Becker is based on SD's maleic anhydride process.

In a brief summary of his experience under the title of a history of Scientific Design, Egbert wrote:

> Egbert contributed the research and engineering talent needed to develop, design and build safe chemical plants. Landau's major contribution was his sales and negotiating ability and his gift of dealing with patent and contract lawyers. Rehnberg's contribution was to give Landau the courage to start the company and in the early years he helped with sales and occupied himself with accounting functions until these were taken over by a professional.

In his later writings, Egbert bitterly resented his treatment and bad-mouthed Landau, as might be expected. Landau minimized Egbert's contributions in his own later writings. Egbert continued his career as a private consultant until his death in 1991 (Fig. 7.5).

The significance of the discovery of the Mid-Century process cannot be overstated. Polyester became by far the most important textile fiber globally, with more and more

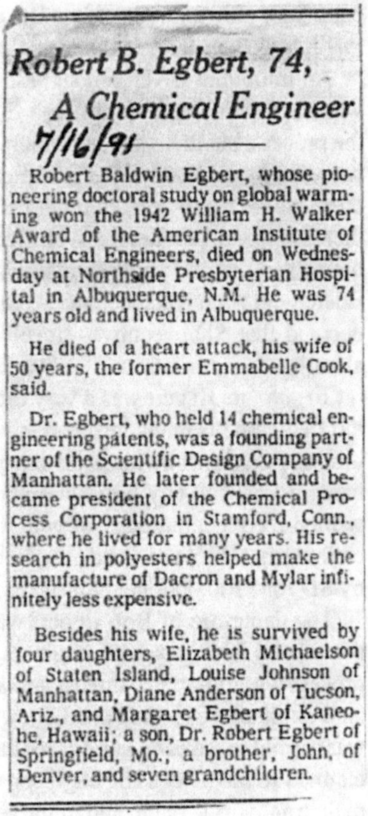

Fig. 7.5 Bob Egbert
obituary

> ## Robert B. Egbert, 74, A Chemical Engineer
> 7/16/91
>
> Robert Baldwin Egbert, whose pioneering doctoral study on global warming won the 1942 William H. Walker Award of the American Institute of Chemical Engineers, died on Wednesday at Northside Presbyterian Hospital in Albuquerque, N.M. He was 74 years old and lived in Albuquerque.
>
> He died of a heart attack, his wife of 50 years, the former Emmabelle Cook, said.
>
> Dr. Egbert, who held 14 chemical engineering patents, was a founding partner of the Scientific Design Company of Manhattan. He later founded and became president of the Chemical Process Corporation in Stamford, Conn., where he lived for many years. His research in polyesters helped make the manufacture of Dacron and Mylar infinitely less expensive.
>
> Besides his wife, he is survived by four daughters, Elizabeth Michaelson of Staten Island, Louise Johnson of Manhattan, Diane Anderson of Tucson, Ariz., and Margaret Egbert of Kaneohe, Hawaii; a son, Dr. Robert Egbert of Springfield, Mo.; a brother, John, of Denver, and seven grandchildren.

plants continuing to be built in China, India, and elsewhere today. Most, if not all of these plants use a version of the SD/Amoco PTA technology for the production of purified terephthalic acid, the main component, with ethylene glycol, of polyester fiber. The patents for producing TPA have long ago expired and the unpatented production knowhow has also been broadly disseminated, resulting in the fact that any new producer can now build a huge TPA plant if he has the funds to do so. But it is fitting to recall that the secret to producing TPA directly in high yields was discovered by Scientific Design's Bob Barker, who reached for bromine, a manganese catalyst and acetic acid to make history.

References

Herbert Meyer (1997) Oral History recorded at Chemical Heritage Foundation
P. H. Spitz 1988 *Petrochemicals. The rise of an industry.* John Wiley & Sons. New York. 286

Ralph Landau and Alfred Saffer. 1968. *Development of the MC Process.* Chemical Engineering Progress Vol. 65, No. 10. 24

Ralph Landau. 1994. *Uncaging Animal Spirits.* The MIT Press Cambridge, MA, 23.

Winfield, J. R. 1953 The *Development of Terylene.* Textile Research Journal. May, 1953

Chapter 8
More Inventions: Nylon Intermediates, Isoprene, Propylene Oxide, Methyl Methacrylate

Abstract SD developed its own approach to research and development, which involved targeting the most attractive research areas, obtaining significant laboratory results and finding a financial partner. Over the next two decades, SD's laboratory and chemical engineers succeed in developing three other major petrochemical processes: synthetic isoprene, KA Oil from cyclohexane, and a co-product route to propylene oxide and styrene. The three technologies found commercial application in the production of nylon, poly-isoprene tire rubber, and polyurethanes and polyester resins. A new route to methyl methacrylate was also under development.

Chapters 6 and 7 cover five processes that illustrate the inventiveness of SD's research and development, three of them leading to substantial financial rewards. Before describing these other technologies in detail, it is instructive to outline the research-oriented and risk-taking atmosphere that prevailed at the company and the skills and approach, largely high level chemical engineering, that led to the firm's success. Ralph Landau has written extensively about this.[1]

Scientific Design Company's top executives quickly became convinced that their firm's success would be strongly favored by three factors, namely:

- That a small, entrepreneurial company could develop technology more quickly than its competitors in large chemical companies.
- That a company created by, managed by and employing mostly experienced chemical engineers would have an important advantage over a company with a hierarchal organization structure, generally using people from various backgrounds and disciplines traditionally involved in a development program.
- That an engineering firm targeting new technology to a global market with many potential takers has a better chance to see its new process commercialized, given typical issues that keep a company from adopting a new technology that would replace a current one.

A paper written by Ralph Landau and Dave Brown (Senior Vice President of Halcon International) for the Institute of Chemical Engineers in London in 1965 substantiates the fact that small companies carrying out research may often have

---

[1]Landau (1994). See Index 414–416.

© Springer Nature Switzerland AG 2019

P. H. Spitz, *Primed for Success: The Story of Scientific Design Company*,
https://doi.org/10.1007/978-3-030-12314-7_8

a considerable advantage over large companies.[2] They cite evidence collected in a series of interviews with R&D managers that suggests several things favoring the small company, notably including the ability of the researchers, and their attitude toward costs, communication and coordination.

It was found that the small company gave the researcher the ability to exert more influence and more independence. Also, small companies tend to hire researchers with experience, while large companies hire researchers directly from schools. And importantly, the small company researcher is more aware of the effect the results of his research will have over the fate of his company. Not surprisingly, it was found that large companies often spend three to ten times as much money as smaller ones to develop a similar product or process.[3]

All of this makes it easier to understand and appreciate the reasons for SD's success. Landau and Brown describe the conditions and atmosphere under which research and development were carried out in its laboratory and engineering office. Research was directed by an experienced officer of the firm with authority and knowledge, the same type of person who may not necessarily have been promoted in large companies, which often tend not to promote scientists and engineers.

John Schmidt, who had a distinguished career in process development and commercialization at SD, contributed a vignette describing Joe Russell, the firm's research head during the time when the key inventions were occurring. "He was the driving force behind the Oxirane technology. One of the many chemical engineers hired by Ralph Landau from MIT, he became SD's director of research at barely thirty. Lean and red-headed, Russell was a little less bantam-sized than Landau, but with much in common—brilliant, driven, visionary. Also detail-oriented, he grilled staff on facts, challenged plans and pushed schedules. Later, after moving from Little Ferry to New York, he held restaurant dinner meetings with his staff: free-wheeling, alcohol-fueled business discussions. They sometimes ran late, once so late that he ended up sleeping on the couch in my Manhattan apartment rather than trekking home to New Jersey. My family took it in stride. He got results and I enjoyed working for him".

Another difference is that in large companies, other departments of the firm (Operations, Sales) often oppose the further development or commercialization of new technologies due to vested interest in existing plants and processes. This did not happen at SD, since there were no existing technologies and there was every incentive to move forward as quickly as possible toward the realization of a new invention. With many companies likely to be interested in the new technology, there was no dearth of potential partners that would allow SD to succeed in a first-of-a kind project. Landau would say, "we search early for companies, anywhere in the world, who might have such an adequately strong incentive that they would be prepared, in the right circumstances, to take the risk of being the first producer using the new process.[4]

---

[2]Ibid. 61.

[3]Cooper A.C. 1964 Harvard Business Review May/June.

[4]Ralph Landau. Op. cit. 64.

Not surprisingly, given the background of its founders, the authors stressed the critical importance of chemical engineers in all phases of its business, listing ten positions occupied by Chem. E's in its organization. The following is abbreviated from the article:

– *Heterogeneous Catalyst Research*: Working with organic chemists, the engineers use their knowledge of kinetics and heat transfer to help design the catalyst.
– *Patent and Legal Work*: Cooperation with patent attorneys results in viable patent applications and effective development of technical and legal details that go into plant design and licensing contracts.
– *Process Development*: Chem. Engineers direct the researchers to develop necessary data for scale-up and economic design.
– *Statisticians*: Chem. Engineers with this type of background are best able to plan and direct experiments.
– *Unit Operations Engineer*: Having the requisite skills in kinetics and heat and mass transfer that go into the design of equipment and processes.
– *Process Evaluation Engineer*: Skill in developing relationships between process variables and costs, carrying out numerous estimates with flow sheets having varying amounts of detail.
– *Sales*: Since SD sales executives are chemical engineers, they can explain the new technologies in economic terms to clients and negotiate technical matters for contracts.
– *Project Managers*: Well suited to form the key link between R&D people, engineering staff and client to optimize and expedite plant design.
– *Startup Technical Directions*: Chem. Engineers have the skill to diagnose and cure operating problems.
– *Administration*: SD's officers are generally Chem. Engineers, as they understand both technical and business issues.

The paper identifies the following questions governing SD's decision to embark on a research program, namely (1) Are the projected economics better than those of current technology? (2) What are the chances of achieving success? And (3) Is the estimated market size and price sufficient to make the new technology attractive to operating companies?

The important inventions covered in this chapter more or less followed the principles described by Landau and Brown and illustrate the ability of SD's researchers and engineers to use their own initiative and the support of their supervisors to arrive at a successful outcome. All of this sounds very logical and well-thought out. But it does not tell the complete story. A letter to Bob Egbert, made available to the author, written by Harry Peters who was most of the time in charge of all engineering at SD, cited the sales philosophy that had worked so well up to the last project, the failed MEG plant. The sales approach was always that a company like SD, with little capital, but major chemical engineering skills, must take the greatest technical risks in order "to get the job", because "no job, no company". The more conservative engineering group countered that the plant must be designed to work, because "inoperable or uneconomic plant….no next plant". It turned out that the engineering people were wrong in the case of SD. While undue process and even mechanical risks and unrealistic cost estimates might be employed to gain the contract, these risks could be substantially reduced during the long periods of design, construction and startup. Peters therefore concluded that the company found itself in the "engineering change business" rather than in the "engineering business". He also said that he always went along, but with one proviso…there could never be a proposal with both an untenable process risk and an untenable mechanical or corrosion risk. Significantly, this concept will be explored in the story of the MEG plant in Chap. 12.

8.1   The Next Breakthrough: High Yield Cyclohexane Oxidation

The production of Nylon 6/6, discovered by Dr. Wallace Carothers of DuPont, involves the reaction of adipic acid, an aliphatic dibasic acid, with hexamethylene diamine (HMDA). Classically, adipic acid is produced by the two-step oxidation of cyclohexane, a ring compound found in refinery naphtha streams. The first step, which produces a mixture (usually called KA oil) of cyclohexanol and cyclohexanone (OL-ONE) is traditionally carried out at elevated temperatures and pressures using manganese or cobalt naphthenate catalysts. Cyclohexane conversion is limited to 10–12% to reduce the formation of undesirable byproducts. Much research had been carried out to improve conversion, but little success had been achieved. In the second step, KA oil is reacted with nitric acid to give adipic acid. In the next step adipic acid is reacted with ammonia to make adiponitrile which is then hydrogenated to hexamethylene diamine (HMDA). This reacts with adipic acid to make Nylon 6/6. KA oil is alternately used to make caprolactam, which is the main ingredient for the production of Nylon 6, the polyamide discovered by German scientists independent of Carothers' work. Caprolactam is made from cyclohexanone oxime via Beckman rearrangement, the oxime being produced from cyclohexanone. The caprolactam is polymerized to Nylon 6.

In the early 1960s, SD management decided that it would carry out original research in the area of nylon 6 or 6/6 components. There was no specific target, except that all possible avenues leading to the two nylons were to be considered. Nylon production was relatively new, started only fifteen years or so earlier as a brand new technology. That would normally mean that there could be an attractive opportunity to develop other, patentable routes to make either or both of the monomers, i.e. adipic acid and HMDA. Adipic had not previously been made in large amounts in the U.S. and the same was true for the other monomer (Fig. 8.1).

DuPont had decided to place the production of both nylon intermediates in the Ammonia Division, which had developed the best process for making nitric acid and depended on this chemical for a substantial part of DuPont's profit at that time. This division had a research department which was undoubtedly disinterested in

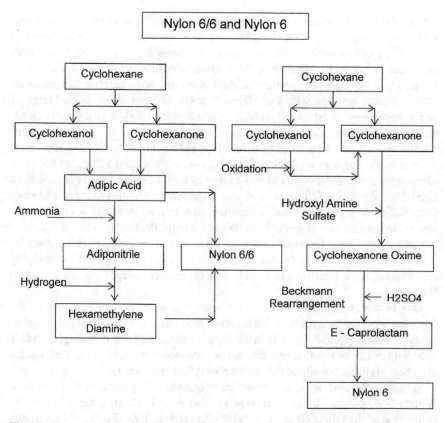

Fig. 8.1 Production routes for nylon 6/6 and nylon 6

exploring other options than nitric acid for the cyclohexane oxidation, being strongly dedicated to using this chemical for the production of nylon. The policy of DuPont of having a number of divisions with widely different skills actually resulted in nylon manufacture to be divided into monomer production (Ammonia Division), polymerization and melt spinning (Chemical Department) and Marketing (Textile Division).[5] The Ammonia Division had no interest in exploring other techniques for oxidizing cyclohexane, given the fact that nylon production had only been carried out for a short period of time and was extremely profitable, "a gold mine in supplying nylon intermediates to the Textile Department."[6] Moreover, some of the brightest DuPont researchers, such as James S. Maloney and Robert L. Pigford, had recently left to head research departments at the universities of Kansas and Delaware.[7] In every respect, therefore, the stage was set for researchers in other companies to

---

[5] Hounshell and Smith (1988).

[6] Ibid. 477.

[7] Ibid. 295.

explore the area of cyclohexane oxidation, using oxygen instead of nitric acid (Later, when environmental considerations became more important and the discharge of nitric off-gases into the atmosphere became objectionable, there would, in any case, have been strong interest in finding other means of oxidizing cyclohexane).

At SD, a team was assembled under John Colton, one of the top chemical engineers with a research bent, consisting of Martin Sherwin, Sherwood Fox, Irwin Margiloff and Joseph Jewett. Sherwin had recently joined SD after receiving a degree in chemical engineering from City College in New York. He recently said, there was a spirit of excitement there and a group of outstanding engineers that made SD a wonderful place to work. As Sherwin recalled, the nylon intermediates research group he joined (there were a couple of others in this area) was an interesting team. Fox had been hired from American Cyanamid and was an expert in thermodynamics and reaction kinetics, Margiloff was highly experienced with hardware, and Jewett not only seemed to have an encyclopedic knowledge of chemicals, but was also an outstanding researcher of the literature at a time when current rapid researching methods (e.g. Chem Abstracts in electronic form) were not available. All three were highly experienced, while Sherwin had no industrial experience except for a summer job at Exxon.

One of the areas investigated by the team was a technology patented by the Japanese firm Ube Industries, which involved production of cyclohexanone by reacting cyclohexane photochemically in the presence of nitrosyl chloride using a 7.5 KW light bulb. Sherwin recalled that this did not produce any useful results, except for deep suntans in the middle of the winter for all four researchers!

Undoubtedly, other approaches were investigated, both by that team and by others. A great deal of research had been done by other firms to improve the selectivity or conversion of the OL-ONE reaction, which proceeds at 100–150 lb per square inch and around 150 °C, with little success.

Some adjuvants such as silicic acid or boric acid had been patented by others for enhancing the oxidation of long chain paraffins. At a chemical conference in Rome, attended by SD researcher Dr. Rex Lidov, N. M. Emanuel a scientist from the Russian Academy Of Sciences, presented a paper in *Tetrahedron* magazine describing the used of boric acid in cyclohexane oxidation, with cyclohexanol being tied up as the borate ester. The results were not considered good enough to be commercially viable but they were very interesting to Lidov, who brought the article back to New York for consideration by SD's brain trust, i.e. Dave Brown, Joe Russell, Al Saffer, and, of course Ralph Landau. The group immediately considered the approach using boron to be very interesting and put several researchers to work to try a different way to carry out the reaction.

After months of work, it became obvious that liquid phase water should not be present in the reaction. Another important finding was that under desirable temperature and water vapor conditions, the ortho-boric acid was dehydrated to meta-boric acid, which allowed the reaction to take place. Then, an idea came up. What if the cyclohexanol sequestered by the boric acid were removed from the reactor, the OL and boric acid separated and the boric acid recycled to the reactor? This sounded like an interesting approach and was immediately tried! It was a Eureka moment! Not only did the selectivity of the reaction greatly increase, but the traditional ratio of OL to ONE could be shifted from the traditional one OL to one ONE to a ratio of 9 or 10 to one. This happened because under that regime OL was removed before some of it could be converted to ONE. The oxidation reaction was, in fact, known to proceed from ANE to OL to ONE to byproducts (some liquid, some solids, which were termed "coffee grounds" by the SD chemists). So, when much of the OL (cyclohehanol) was removed when it was formed, the combined selectivity to Ol and ONE was greatly increased and the amount of solid byproducts formed was greatly reduced, raising the yield significantly, from 65 to 75% in the classical system to 90–95% using boron.[8]

Many experiments carried out over a number of months confirmed this desirable way to carry out the cyclohexane oxidation. The SD researchers knew that they had the start of a new process. The reaction system could be designed to carry out the oxidation in one vessel, continuously withdrawing the borate ester to separate out the captured cyclohexanol and returning the boric acid to the reactor, somewhat similar to the way a fluid bed cat cracker works. As reported by Landau and Dave Brown in a paper presented at an AIChE symposium in New Orleans in 1963,[9] the engineering of the process required more than normal experience and ingenuity since three phases—solid, liquid and vapor—are handled simultaneously throughout most of the process. Nevertheless, SD's engineers were up to the challenge. The work carried out in the small pilot plant was sufficient to allow the development of a workable flowsheet, as was recently recalled by Neil Yeoman, who headed up the design of several cyclohexane oxidation plants in the coming period.

A number of patents were obtained on the process, but then SD hit a snag. Consternation developed when it was found that an inventor, sometime before, had obtained a broad patent covering the use of boric acid in various types of oxidations, and this would cause an infringement problem for SD's technological breakthrough. Now came a fascinating development. It turned out that the inventor, a private individual, had obtained the patent because he had carried out research on an ingredient that went into a consumer application, allegedly a hair treatment product, which he had decided to protect with a patent. What happened next was recalled by the author who, at that time was working for Landau. SD's management decided it would attempt to purchase a sublicense from this inventor just for practicing its nylon intermediates process. SD hired a consultant, Bob Purvin, to meet the inventor and to negotiate a sublicense. The consultant contacted the inventor and they had a single meeting. When Purvin returned, he said that they had agreed on five thousand dollars, to the delight of the inventor, who saw it as "found money" For SD it was obviously a small price to pay for becoming free to license its new process to nylon producers (Fig. 8.2).

---

[8]Ralph Landau. Op. cit. 34.
[9]Ralph Landau. op.cit. 32.

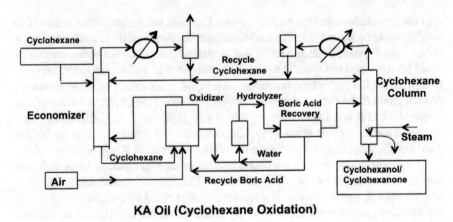

KA Oil (Cyclohexane Oxidation)

Fig. 8.2 Cyclohexane oxidation process with boric acid sequestering cyclohexanol

The first opportunity for commercial application turned out to be in Australia, not to make adipic acid, but rather because Monsanto was interested in building a small phenol plant. SD worked out a scheme to use the new technology as follows: With no cyclohexane available, benzene was hydrogenated, using classical catalysis. Marshall Frank was asked to review the literature to come up with a suitable design. The cyclohexane was oxidized to the mixture of OL and ONE, the cyclohexanol was separated and was subsequently dehydrogenated to phenol. This seems a little complicated, but it worked. Unfortunately, in an attempt to keep down the cost of the plant, no intermediate storage had been provided between the stages. The reaction sequence was too complex to operate flawlessly in sequential mode. Addition of intermediate storage would probably have solved the problem. Nevertheless, the individual steps worked well and the new cyclohexane oxidation technique was proved out successfully, giving SD some feedback for designing subsequent plants.

The plant was operated for a short time and then shut down and converted to other uses. Two years later, Monsanto Australia sued SD for non-performance, asking $10 MM in damages. SD defended itself, citing a legal term called *laches* and *estoppel,* which is applied to a situation when a company does not allow the other party to rectify the situation that caused the damages. The dispute went all the way to the U.S. Supreme Court, which ruled that the issue should be solved by arbitration. Eventually, Monsanto and SD settled the suit when SD allegedly granted Monsanto a royalty-free license to use the cyclohexane technology to make adipic acid in its large nylon plant. (There was also a rumor that SD had given Monsanto a ten percent share in Scientific Design Company, but this was never confirmed). That plant had been built by Monsanto's subsidiary Chemstrand Corporation after receiving a license from DuPont under an antitrust decree shortly after the war. Under the terms of the agreement, Chemstrand's adipic acid plant was converted from the traditional nitric acid oxidation process to the new SD technology.

No other phenol plants were built using the SD process, though SD had publicized it widely.[10] Around the same time, SD licensed the cyclohexane technology to a British firm, Howards of Ilford, who used a batch process to make the alcohol and ketone. Little seems to be known about this plant.

....

SD's new technology was of great interest to the world's nylon producers, both those producing Nylon 6/6 (from adipic acid) and Nylon 6 (from caprolactam). The firm designed, engineered and started up seven additional commercial plants for the production of KA Oil. These were for Imperial Chemical Industries (Two plants), LaPorte Industries, Rhone Poulenc, Farbenfabriken Bayer (two Nylon 6 plants), and Mitsubishi Chemical Industries. The new technology was attractive due to substantially lower raw material and utility costs, with only slightly higher capital costs versus the conventional technology. The new cyclohexane oxidation process had substantial economic advantages over then commercial nylon and caprolactam processes. The molar yield of cyclohexanone plus cyclohexanone, 90–95%, was substantially higher than the traditional yield and the cost of utilities was about half of current technologies. The capital cost of the SD process was, however slightly higher (Fig. 8.3).[11]

One other application of the new process came about. In the mid-1960s, when DuPont was developing a new polyamide textile fiber based on cyclododecane (12 carbon atoms), it swallowed its pride and turned to SD to ask whether its technology could be used for the oxidation to produce dodecanedioc acid in the next step. Analo-

Fig. 8.3 SD's cyclohexane oxidation process used by ICI in its nylon 6/6 plant in England. *Source* AkzoNobel

---

[10]*Recent Developments in Industrial Chemistry.* (1961) Ind. and Eng. Chem. Vol. 53, No. 10. Oct. 1961 37A.

[11]Ralph Landau. Op. cit. 34.

gous to the production of adipic acid. DuPont received an SD license and carried out the necessary research and piloting. It then built a plant and started the production of Q*iana*, an especially "silky" nylon-type fiber used for fancy shirts and blouses. The company had already spent $68 million on R&D and invested $120 million in the plant, including oxidation, polymer production, and fiber spinning. Unfortunately for DuPont, it turned out that the Japanese had also developed and started marketing a fancier textile fiber, but produced from cheaper polyester. Sales of Qiana topped out at $79 million in 1977, then dropped off. DuPont stopped production in the early 1980's with cumulative operating losses in excess of $200 million.[12]

8.2 Isoprene

Although styrene-butadiene rubber was found to be a very suitable material for production of tires, it did not fully measure up to natural rubber, both in terms of physical properties or because of price. Natural rubber has a greater resistance to heat and this made longer-lasting tires. Butyl rubber was used to make the inner tubes. So, there was no all-purpose synthetic rubber. 1–4 poly-isoprene is a synthetic rubber built in a certain geometric form and is therefore known as a "stereospecific" rubber, almost indistinguishable from natural rubber. In the 1950's, research work commenced in several companies to find a route to make isoprene, a five-carbon unsaturated molecule that can be polymerized to poly-isoprene. While isoprene was already an article of commerce, it was very expensive. It would have to be made at a cost similar to that of butadiene. The challenge therefore was to develop a route to isoprene that had (a) low capital investment, (b) an abundant, low cost raw material, (c) high yield and (d) a pure product.

Poly-isoprene could be used in admixture with natural rubber and would be useful if for some reason natural rubber might become too expensive or in short supply. Isoprene is a five-carbon double unsaturated molecule similar to four-carbon butadiene and has a molecular structure closely resembling that of natural rubber. Isoprene became available as a valuable byproduct when naphtha and gas oil cracking supplied copious amounts of byproducts. However, many Gulf Coast crackers could switch among feedstocks almost at will. This meant that the amount of byproduct isoprene could drop sharply from month to month as cracker operations shifted to light hydrocarbon feedstocks. Goodyear Tire and Rubber Company had for some time produced poly-isoprene rubber and was therefore concerned that its availability was at the mercy of cracker feedstock choice. And so Goodyear wondered whether isoprene could be produced in an on-purpose plant at a competitive cost.

Goodyear and SD looked at a number of possible routes in the literature. First, there was a process that reacted methyl ethyl ketone (MEK) with formaldehyde, followed by hydrogenation and dehydration to form isoprene. This was eliminated due to the high cost of MEK. Next, there was a possible acetone-acetylene route, which

---

[12]Hounshell. Op. cit. 439.

looked more economical but still had excessive raw materials cost. The next possibility considered was an isobutylene-formaldehyde synthesis, but again, economics looked unattractive. The researchers concluded that only a refinery product-based route would be attractive economically. This led to consideration of isopentane or isoamylene dehydrogenation. Here, Goodyear concluded that refining firms would be in a better position than a rubber company like Goodyear, if they wanted to use their own feedstock to make isoprene. This eventually led to the decision to try and develop an isoprene process based on propylene.[13] Goodrich asked SD to set its research team on a quest to develop an isoprene-from-propylene process.

SD's researchers set out to meet this challenge. They proposed that a good way to make a molecule with five carbon atoms was to dimerize propylene, which gives an unsaturated five atom straight chain molecule with a methyl group side chain and to then knock off the side chain. Dimerization with different catalysts gave products that contained a large number of possible isomers including methyl-pentenes, ethyl butenes, and dimethyl butenes. A number of different catalysts and processes were studied and finally one was found that was specific enough to give substantial amounts of the desired methylpentene isomers. Trialkyl aluminum catalyst gave efficiencies of about 95%. The inconvenient fact was that the double bond resulting from dimerization was at the end of the chain and therefore in the wrong position to produce isoprene. A catalyst was needed to isomerize the 2-methyl-1 pentene to 2-methyl-2 pentene. Since that would give a reactor effluent that had both the converted and unconverted dimer, this would, of course, involve a major recycle operation, requiring the separation by distillation of the two isomers and recycling the unconverted isomer. The final step would be to crack the desired isomer at 650–800 °C, knocking off the side chain and yielding isoprene.[14]

Concentrated research work eventually yielded the desired three step process. The pentene separation step turned out to be the most difficult due to the very close boiling points of the two isomers. "Calculations covering this separation was a bear", was recently recalled by Martin Sherwin, a member of the team. "In those days we didn't have a computer and had to use hand calculators to do the math". As to the cracking step, this was to be a bromine and acetic acid-based process, somewhat similar to that employed in the terephthalic acid technology described in the previous chapter. Eventually, a pilot plant was built and the process was ready (Fig. 8.4).

Goodyear built a thirty million pound per year plant at Beaumont, Texas at its large synthetic rubber complex. The two first steps worked well, but in the cracking step a lot of solids were formed and a substantial amount of corrosion occurred. The plant was shut down to decide what should be done. Earlier lab work had shown that ammonium sulfide worked well instead of bromine, though the yield of isoprene was somewhat lower. But the reaction was obviously much less corrosive. This led to some redesign of the plant and a new startup. This time, the entire process went well and SD had produced pure isoprene, well suited for making polyisoprene. Brian Ozero, who had recently joined SD, was assigned to do the redesign, using some

---

[13] Anhorn et al. (1961).
[14] Anhorn et al. (1961).

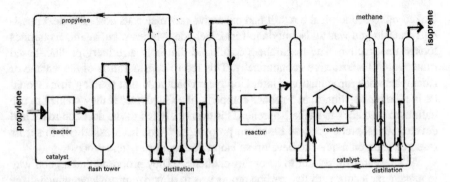

Fig. 8.4 Goodyear-SD's three-step route to isoprene. *Source* chemical week

new and some of the existing equipment. He told the author that he decided to come to work at SD after working at other jobs and being told by a senior Shell manager that SD was now the most exciting place for a young chemical engineer to work. So here a relatively new "recruit" had a significant part in making an interesting process work.

Success in developing the isoprene process led SD to look for other customers and so a team visited a number of countries in different parts of the world offering the isoprene technology. There were, apparently, no takers. Ozero recalled that around that time, polypropylene demand and production was surging, putting strong upward pressure on propylene, which until then had refinery (close to fuel) value. The economics of isoprene production from propylene no longer looked attractive.

Unfortunately, the Goodyear plant had a fire and needed major repairs. Also coincidentally, ethylene plants at that time were cracking large amounts of heavy liquids co-producing substantial quantities of isoprene, whose price sank to very low levels. And so, considering the higher price of propylene and the availability of relatively inexpensive cracker byproduct isoprene, Goodyear decided not to rebuild the isoprene plant, a monument to successful research and development but a victim of economics.

8.3   Epoxidation of Propylene

It was inevitable that the researchers at Scientific Design would target propylene oxide as an opportunity to develop a direct oxidation process, recognizing that many others had tried and failed. The market for propylene oxide was growing rapidly with primary applications for polyether polyols going into urethanes and propylene glycols into polyester resins. Propylene oxide, similar to ethylene oxide had been produced for a long time via the chlorhydrin process, though Union Carbide had developed a direct process for ethylene oxide in the 1930s, as described in Chap. 6 and SD now also had a direct EO process. At the time SD was carrying out its research there were several U.S. producers making propylene oxide using chlorohydrin. For Dow Chemical, this was no problem, since it was a chlorine producer and could recycle the propylene dichloride byproduct of the chlorohydrin process to its hydrocarbon chlorination furnace where the chlorine values were recovered. Jefferson Chemical, a joint venture of Texaco Chemical and American Cyanamid that also produced propylene oxide "the old fashioned way" did not have a similar situation and was therefore a high cost producer. Olin also had a propylene oxide plant. It is not known whether any of the then current propylene oxide producers had been doing any promising research on a direct route. In Europe, Bayer and BASF were major producers of polyurethanes. BASF may have been active in propylene oxide research, but only decades later developed a direct process. In sum, there was currently no way to produce propylene oxide except by the chlorhydrin route and this was the case both in the U.S. and in Europe, where Bayer had invented polyurethanes.

The history of unsuccessful research on direct routes to propylene oxide had undoubtedly been studied carefully by the SD team assigned to develop such a process. These attempts included non-catalytic vapor phase oxidation (maximum selectivity: 25%), liquid phase oxidation (selectivities in the 30–50% range with substantial amounts of formic and acetic acid) and a process based on acetaldehyde, which made a large amount of acetic acid co-product).[15]

A great deal of time and effort was spent on trying different approaches and catalysts for improving the yield in a liquid phase process. Eventually, this led to using hydrogen peroxide, which proved to be quite unsuccessful. Martin Sherwin (see earlier in this chapter) recently said that after four or five unsuccessful tries with hydrogen peroxide he somehow screwed up his courage to tell Ralph Landau to forget using this oxidant as its use would certainly prove to be uneconomical! Whether this was instrumental in SD shortly switching its research to the use of organic hydroperoxides will never be known. However, when this was tried, success was immediate. From this point on, all furtrher work was on co-product processes, prominently on co-producing styrene with propylene oxide.

Interestingly, some decades later, Dow and BASF commercialized a direct propylene oxide process based on hydrogen peroxide.[16] Evonik in Germany also developed a process of this type.

SD now had the beginnings of a process that would always involve the production of a co-product together with the desired propylene oxide. Moreover, the amount of co-product would be of the order of twice the amount of propylene oxide produced. The concept of developing a process that would have a valuable co-product together with propylene oxide was not new. Not long before, Hercules perfected a co-product process for phenol, based on benzene and propylene that co-produced significant quantities of acetone. This technology almost totally replaced an earlier phenol process based of chlorine, signifying that a co-product process could be viable if the co-product has a large market. With this background, SD felt safe in developing a process of this type, provided the co-product was eminently marketable.

---

[15]Ralph Landau Op. cit. 40–41.

[16]*Propylene Oxide Routes take off.* 2009 Chemical & Engineering Progress. Oct. 9, 2009 22–23.

The process SD then started to develop was based on the original discovery by SD's research director Joseph Russell and researcher John Kollar that hydroperoxides can react with olefins at 120–140 °C and around 500 lb per square inch in the presence of catalysts containing molybdenum, vanadium or titanium to give high yields of both alcohols and epoxides. With propylene, selectivities to propylene oxide can be better than 95% on a molar basis.[17] Epoxidation with organic hydroperoxides was the key step. SD studied and patented different organic hydroperoxides (e.g. ethylbenzene, cumene, others) and a broad range of soluble catalysts for the epoxidation (e.g. molybdenum- titanium-, vanadium naphthenates). The reaction was carried out at 250–1000 psig and 80–130 °C.

The production of propylene oxide proceeded as follows. In the first step, the peroxidation of ethylbenzene produces the organic hydroperoxide using conventional technology at very high selectivities. Following that, in the peroxidation reaction, propylene and the hydroperoxide are mixed with the catalyst from 1 to 3 hrs. until 98% of the hydroperoxide has reacted. Propylene to propylene oxide selectivities are in the range of 95%, with methyl benzyl alcohol as the other product. This alcohol is then dehydrated to styrene over a supported titanium catalyst in a well-known manner. Pure styrene is obtained as a co-product (Fig. 8.5).

The SD engineers had never encountered a process of this type before. The next step was a small pilot plant, which was operated successfully, giving similar results than the laboratory. Meanwhile, a preliminary flowsheet was developed and equipment design was commenced. Economic studies showed that the process would be

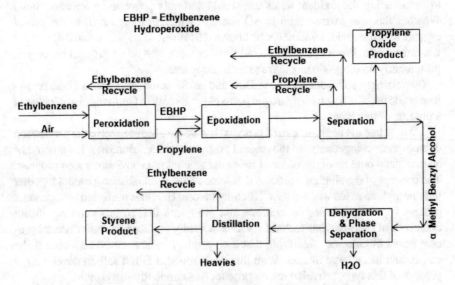

Fig. 8.5 Co-product process diagram for production of propylene oxide and styrene

---

[17]Ralph Landau. Op. cit. 41.

economically much superior to the conventional chlorohydrin technology then in use globally.

In 1963, Landau had believed that it was time to prepare for Scientific Design to enter the area of chemical manufacture. Accordingly, a new company, Halcon International Inc. was set up to be the holding company for a research arm, an engineering company (SD), a catalyst manufacturing company and a future operating company.

It was now time to look for a development partner. Landau later admitted that several companies were informed of SD's invention and interest in a partnership. Then lightning struck when SD's patent attorneys found that Arco Chemical had been developing a similar approach to propylene oxide production, except only with aliphatic hydroperoxides. It appeared that SD would have a better patent position than Arco, but Arco could carve out its own process. It would clearly be preferable if the two companies combined their technologies and became partners. This led to the formation of the Oxirane Company. How the discovery of the unique epoxidation technology led to a complete transformation of global propylene oxide production is covered in Chap. 10.

8.4 Methyl Methacrylate

This chemical was first synthesized in Germany before World War II and became an important product for the manufacture of clear plastic materials (e.g. *Plexiglass, Lucite*) and paints by companies like Rohm and Haas and DuPont. The classical synthesis of methyl methacrylate (MMA) involves the sodium hydroxide-catalyzed condensation of acetone and hydrocyanic acid (made from acetylene) to give acetone cyanohydrin. This undergoes methanolysis to produce MMA. This process is still in use today. Other processes have been under study and development in the United States and Japan, but none of these have replaced the conventional route to any great extent.

In the late 1960s, Joe Russell, then head of SD's research arm, patented a process for the production of MMA via the oxidation of isobutane. The reaction proceeds to t-butyl hydroperoxide which is, in turn, reacted with methacrolein and added alcohol to form methacrylic acid or ester and t-butyl alcohol, with some methacrolein formed and recycled. A later version of this process that was worked on in the 1980s first converted isobutane to isobutylene which is oxidized to methacrolein. Halcon employees recently contacted believed that this process was considered very promising, but was never prioritized to the point where it would be discussed with potential company sponsors in the usual manner. In his memoirs, Neil Yeoman wrote "The Methacrylate technology to which Landau refers in his 1978 speech to the Newcomen Society was said to be as impressive as earlier technologies, but it never got off the ground. By the time it was ready, the failure at Channelview dominated Halcon's thinking and methacrylate never had a chance".

References

V.J. Anhorn, K.J. Frech, G.S. Schaffel and David Brown 1961 *Isoprene from Propylene* Chemical Engineering Progress Vol 57(5) May 1961.

V.J. Anhorn et.al. 1961. *Isoprene Process Evaluation* Chemical Engineering Progress. Vol 57(5) May 1961 11–13.

Hounshell, David A. and John Kenly Smith (1988) *Science and Corporate Strategy. DuPont R&D, 1902–1980.* Cambridge University Press. New York p 262.

Ralph Landau. 1994 *Uncaging Animal Spirits.* The MIT Press. Cambridge, Mass. See Index 414–416.

Chapter 9
Other Firms: A Period of Breakthrough Inventions

Abstract Other significant petrochemical process developments occurred during the time SD was making important inventions. These include UOP Platforming, Monsanto's carbonylation process for acetic acid, acrylonitrile from propylene, Shell Chemical's SHOP process for alpha olefins and linear alcohols, vinyl acetate from ethylene, and ethylene oxychlorination. Established engineering companies were not targeting the development of new, complex technologies, because they did not have experienced researchers and did not want to compete with their industry client base.

While Scientific Design Company was developing a number of novel petrochemical technologies, researchers in other companies also investigated opportunities to invent patentable processes, based on the new highly reactive feedstocks now plentifully available. They looked at current methods of production for important chemicals to see whether a different, more economical route might be developed. They wondered whether a petrochemical feedstock could replace a natural material in a new synthesis or whether it might be possible to eliminate an undesirable byproduct. They experimented with new catalyst systems that might lead to a completely novel process. But companies are not always supportive of research that does not fit with current product lines and it will never be known whether research by motivated scientists in chemical firms like DuPont, Union Carbide, Dow and others may have been on the route of discovering promising new technologies that were not carried further due to lack of funding. Nevertheless, some did meet with success, as covered in this chapter. In some cases, these processes were, in fact, in areas not related to the product lines of the companies they worked for, suggesting that the research work was carried out when a chemist saw an opportunity that might not necessarily fit the company's strategy.

To give credit to these researchers, this chapter will cover six breakthrough technologies commercialized during the period when SD made its important technological inventions. The criterion used to select these processes was that "they changed or created an industry". The technologies highlighted were:

– UOP's *Platforming* process for BTX aromatics
– The Monsanto process for acetic acid via carbonylation of methanol
– The Sohio process for acrylonitrile from propylene

© Springer Nature Switzerland AG 2019

P. H. Spitz, *Primed for Success: The Story of Scientific Design Company*,
https://doi.org/10.1007/978-3-030-12314-7_9

- Ethylene Oligomerization and Shell's SHOP process
- The National Distillers and Chemicals process for vinyl acetate
- Ethylene oxychlorination for vinyl chloride.

Industry observers might argue that other technologies might fit the above description, but the ones listed above stand out as particularly important. To be faIr, it would be appropriate to add the postwar development of high density polyethylene and polypropylene as products that "created a new industry". However, SD's research did not include the development of novel polymers, while the six technologies identified above are in areas that SD might conceivably have been interested in exploring.

Platforming and Monsanto's carbonylation technology, were widely licensed. In the first case, UOP is mainly in the business of process licensing. In the case of Monsanto, while itself an acetic acid producer, it was not interested in trying to dominate the market with its new technology and instead decided to monetize its successful research by licensing the technology broadly to domestic and international companies. The Sohio process was licensed because Sohio also decided to opt for very substantial license fees in addition to entering the industry and becoming a large producer. Shell did not license competitors with its SHOP process, because it was competing with other petrochemical technologies involving linear olefin technologies (Gulf), synthetic detergent alcohols (Ethyl) and with detergent alcohols made from a natural raw material (fatty acids from palm or coconut trees). National Distillers' decision was similar to Sohio's in most respects. Ethylene oxychlorination also fits into this category, since its almost universal adoption changed the global production of vinyl chloride from ethylene.

Scientific Design, during the period covered by these six very important technologies from six different companies, developed five technologies based on its own original research.

- The SD ethylene oxide/glycol process was widely licensed internationally and continues to this day as a leading technology.
- The Mid-Century process (invented by SD) for terephthalic acid became the standard technology for the production of polyester fiber and solid resin, allowing Amoco Chemical to become the dominant worldwide producer.
- The SD cyclohexane oxidation technology, when commercialized, was seen as substantially superior to the traditional nitric acid oxidation technology originally developed by DuPont. It was licensed to a number of the world's largest nylon and caprolactam producers.
- The co-product propylene oxide process commercialized by Arco Chemical and SD in a venture called Oxirane Chemicals was highly successful as a replacement for the traditional chlorhydrin technology. The joint venture came about when SD and Arco Chemical found that both companies had researched similar approaches to propylene oxide production, with SD patenting styrene as a co-product and Arco patenting butyl alcohol. The two partners then assigned both technologies to Oxirane.

– The SD isoprene process, using very unusual chemistry, was successfully commercialized by Goodyear Tire and Rubber Company. A dramatic lowering of byproduct isoprene prices made this breakthrough technology uneconomical.

Of the six important technologies identified above and commercialized by companies other than Scientific Design, one was by an engineering company with a strong research background and capability: Universal Oil Products (UOP). Like SD, UOP was in the business of developing new technologies and licensing them, in UOP's case, to as many global operating companies as possible, given its background in developing licenseable technologies for the petroleum refining industry. When the petrochemical industry offered opportunities, UOP became very interested in joining this game. In fact, it developed or co-developed several other petrochemical technologies, notably p-xylene separation from its isomers (Parex technology), linear paraffin separation for detergent applications, and some others. UOP did not get involved in oxidation technologies, which was a key area for Scientific Design.

Another engineering firm, Institut Française des Petrôles (IFP), headquartered in Paris, similar to UOP started as a research and engineering firm specializing in refining technology, but it later also turned its attention to petrochemicals. IFP gained some success in developing and offering some petrochemical processes, but these tended to be similar to existing processes with some wrinkles to get around patents.

As discussed in Chap. 6, engineering firms like M. W. Kellogg, Badger, Foster-Wheeler, and Stone & Webster did not emulate Scientific Design in identifying opportunities for petrochemical process development. These companies were well established, some with competent research chemists and experienced chemical engineers. The reason, most likely, is that these firms were loath to compete in research with their operating company clients, different from SD which initially had no clients and had no problems with the concept of how their research might be competing with the research activities of their potential client base. SD just wanted to be ahead of them. Firms such as Kellogg and Badger, on the other, hand did little research on chemical syntheses, but mainly looked for opportunities to differentiate themselves from their competitors. As an example, Kellogg partnered with ICI to offer a new large ammonia process. Badger became known for its expertise in fluid bed technology. Several developed their own version of giant naphtha crackers for ethylene production, claiming advantages in cracking furnace and quench boiler design and in other parts of the ethylene production flowsheet.

9.1 UOP Platforming

Vladimir Haensel was born to Russian parents in 1914 where his father was a professor of economics. Escaping from the U.S.S.R in 1928, the family settled in the U.S. in 1930, where Haensel's father became a teacher at Northwestern University. Haensel gained his science degree at Northwestern and earned a master's degree in chemical engineering at MIT, joining UOP in 1937.[1] He was assigned to the High Pressure Laboratory at Northwestern as an assistant to the famous catalyst researcher Prof. V. N. Ipatieff, who had been a professor of chemistry at St. Petersburg University and now consulted for UOP. Working on various types of catalysts, Haensel continued his studies and gained a Ph.D. degree in chemistry from Northwestern. Among other areas, he worked on antiknock properties of various hydrocarbons in his early years.

By the late 1930s, the octane number of gasoline could be improved considerably by a number of ways, including alkylation of propylene and butylenes or addition of tetraethyl lead. The problem was that naphthas going into the gasoline pool, even when subjected to catalytic hydroforming with acid catalysts, did not exceed a 65 Research Octane number measurement, which was too low to keep engines from knocking and could not be "leaded up" to a desirable level with TEL.

In the mid-1940s, UOP spent a lot of time studying the hydrocracking and aromatization of paraffins and naphthenes. Paraffins are dehydrogenated to olefins which then cyclize to aromatic rings. Naphthenes are dehydrogenated to aromatics. It was found that the required catalyst to transform low octane napthas to a more desirable form had to have two different functions, namely hydrogenation/dehydrogenation and acidity. Much of the work was directed toward avoidance or minimization of coke formation at the very high temperatures when these reactions take place.[2]

Haensel thought that platinum, a precious metal, should be tried, supported on alumina catalyst. Actually, platinum and nickel had, in fact, been found useful in aromatization reactions, but this was more complicated. At first concerned because of platinum's high cost, Haensel believed that the platinum could be regenerated and reused and that it might be possible to use only small quantities. Experiments proved successful and it was found that the alumina isomerized to aromatic rings the unsaturated hydrocarbons formed by the platinum. However, the catalyst died after a short period of time, due to excessive coking. It was decided to raise the temperature sequentially until it was 200 °C higher than what was thought the platinum could stand. The results were now excellent with high yields and high octane numbers. And there was no coking! The process had another big advantage over conventional naphtha hydroforming in that it generated large amounts of hydrogen. This could then be used to desulphurize impurities in the naphtha fed to the aromatization process.

Attention was now paid to the high cost of the platinum. Initially 3% platinum on alumina, the catalyst was gradually changed to 1.3, then to 0.7%. Also, the catalyst support was changed from aluminum nitrate to aluminum chloride. In order to take advantage of the dual-function nature of the catalyst, the units were designed with three reactors in series. In the first reactor, much of the dehydrogenation and dehydroisomerization of naphthenes and paraffins took place. The two other reactors presumably with a moderately different catalyst, concluded the complete reaction. External reheat brought the inlet temperature of each reactor back to 930 °F.

By varying reactor pressure above or below 700 lb per square inch, the reaction conditions could be arranged to fit the end-product requirements, depending whether

---

[1] Haensel (2006).
[2] Spitz (1988). 176

the naphtha would be used as gasoline or for the production of BTX (benzene, toluene, xylenes) aromatics. The process, called *Platforming*, dramatically raised the octane number of the entering naphtha. A commercial test run was carried out at Old Dutch Refinery at Muskegon, Michigan in 1949. Company performance data showed that the naphtha's research octane number had been raised from 52.5 for the naphtha fed to the reaction system to 81.3 (unleaded) for the effluent and from 70.2 to 93.9 with 3 cc TEL.[3] By 1958, UOP had installed 106 platforming units with a total capacity of 755,000 barrels per day. *Platforming* became one of the most important processes in the history of petroleum refining. Because it led to the construction of hundreds of naphtha reforming units designed primarily for the production of BTX aromatics, one of the main cornerstones of petrochemicals manufacture, it is included in this list. For his landmark discovery, Vladimir Haensel received the Perkin medal, was inducted into the National Academy of Engineering, and received the National Medal for Science from President Nixon in 1973.

9.2 Monsanto's Acetic Acid Process

Acetic acid is an important organic chemical that is used for the production of vinyl acetate, cellulose acetate, butyl and isopropyl acetate esters, acetic anhydride and other chemicals. Vinyl acetate became a rapidly-growing derivative in the 1960s and, as discussed elsewhere in this chapter, is based on the reaction of ethylene and acetic acid. Monsanto, Celanese, and National Distillers and Chemicals (NDCC) became the three largest producers of acetic acid. This chemical had over the years been produced in a number of ways. Before Monsanto's breakthrough process, acetic acid used in industrial processes was mainly manufactured from acetaldehyde, in many cases using a Hoechst process. The reaction proceeds via peracetic acid, which reacts with acetaldehyde to form two molecules of acetic acid. A yield of 95–97% is achieved. Other companies, such as Bayer and Celanese developed and commercialized acetic acid manufacture using hydrocarbons such as n-butane or n-butene. British Distillers developed technology to oxidize light gasoline fractions. All these latter processes give co-products such as formic and propionic acids, while Celanese also co-produced methyl ethyl ketone and acetone. None of the processed described above were direct, but proceeded via intermediates. BASF for a long time worked on techniques to carbonylate methanol to acetic acid, eventually in the presence of a cobalt-iodide catalyst.[4] Here, again, a number of byproducts were formed. The selectivity to acetic acid was 70%. The reaction was very corrosive, needing the use of very expensive Hastelloy, a molybdenum-nickel alloy. BASF built the first small plant in 1960. Later, as the technology was optimized, two large plants were built,

---

[3]Ibid 182.
[4]Green and Witcoff (2003).

one for BASF in West Germany and the other for Borden Chemical in Geismar, Louisiana.[5]

Monsanto, in the post-war period, became a large producer of oxo alcohols, a process involving low pressure hydroformylation of olefins. At a company conference, a research director said that the price of methanol was coming down and wouldn't it be nice if we could come up with a low pressure process for carbonylation of methanol to acetic acid. At that time, Monsanto had a corporate research organization funded outside of division budgets and managed by Richard S. Gordon. It was charged with inventing new products and processes that could have a large impact on Monsanto. A group headed up by James Roth decided on their own to work on methanol carbonylation in their laboratory. Roth, who managed the Catalyst Research Department, always described his group as carrying out "frontier research", as recently mentioned to the author by Bill Lewis, who spent most of career with Monsanto.

Within eleven days they discovered that a rhodium carbonyl iodide catalyst gave good yields of acetic acid.[6] Roth later explained that while Monsanto was not the first in rhodium-based carbonylation, it was among the first in so-called rhodium arylphosphine complexes for carbonylation chemistry. Roth claimed later that his group had tried to get the Organic Division to try this chemistry on oxo alcohol manufacture but, being risk-averse, turned down this idea in favor of licensing a French process, using an old cobalt system.

The new rhodium-based acetic acid process with iodine promoter worked at much lower pressures than the BASF route, around 500 lb per square inch and 180° Centigrade and achieved selectivity's close to 99 percent(!). The process used a soluble homogeneous catalyst containing an extremely expensive precious metal, rhodium. There was concern that even a 0.0001% loss of catalyst during stream recycles would render the process uneconomic. However, "the process was engineered so exquisitely that no rhodium is lost", according to Dr. Jeffrey Plotkin, a colleague of the author who for many years headed up Chem Systems' Process Evaluation and Research Planning (PERP) program (Fig. 9.1).

The Monsanto engineers scaled up the process in a pilot plant and the company shortly thereafter decided to build a 300 million pound per year acetic acid plant at its complex in Texas City, Texas. Monsanto then decided it would license this technology both domestically and globally.

James Roth was trained as a chemist, receiving his PhD in physical chemistry at the University of Maryland. In 1988 he received the Perkin Medal and was granted other rewards between 1986 and 1991.

An interesting sidelight is that Roth did receive a $7000 reward from his employer for the successful invention of the so-called Monsanto acetic acid process. He later wondered, without rancor, whether that was adequate compensation for generating two hundred million dollars in royalties for the company.[7]

---

[5]Weissermell and Arpe (1978).
[6]Roth (1995).
[7]Ibid.

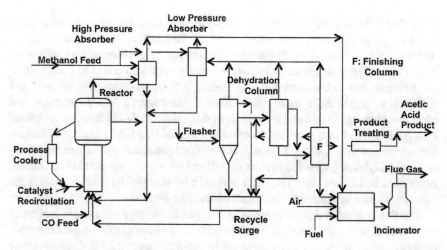

Fig. 9.1 Monsanto's acetic acid process

9.3 Sohio Acrylonitrile Process

First synthesized by IGFarben and used for the German synthetic rubber program, acrylonitrile was a relatively unimportant chemical until sometime well after the end of World War II, when Bayer and DuPont started producing acrylic fiber (*Orlon* by DuPont) which became an important textile material. Not much later, acrylonitrile became an ingredient in acrylonitrile-butadiene-styrene (ABS) resins and a monomer for polyacrylonitrile that was converted to carbon fibers.

The traditional synthesis of acrylonitrile was based on the catalytic addition of hydrogen cyanide (HCN) to ethylene oxide, followed by dehydration of the ethylene cyanohydrin to acrylonitrile. The German process was described in the C.I.O.S. reports referred to earlier in this book and, in the U.S., was practiced by Union Carbide and American Cyanamid. Another process, developed both in Germany and the U.S. was the catalyzed addition of HCN to acetylene, practiced by Monsanto and, later, American Cyanamid.

Allied Chemical scientists in 1949 had studied the conversion of propylene and ammonia to acrylonitrile with different catalysts, but results were not promising and the work was discontinued. In the late 1950s several other companies, including Distillers in England and Sohio in the U.S., also conducted research on the catalytic oxidative reaction of propylene with ammonia.

James Idol earned a chemistry degree at William Jewell College, an M.S. degree in chemistry at Purdue, with a minor in chemical engineering and a PhD also at Purdue. Offered a job at several large oil companies, he chose Standard Oil Company, Ohio (Sohio), starting in 1955, because he felt that he could make an impact at that smaller firm.

At that time, he and his team were studying acrylates and glycerin production based on propylene. Acrylates were then made from acetylene, but it was thought that acrylic acid could be made from propylene via acrolein. Some promising work by Dr. Hans Reppe in Germany had shown that acrylic acid could be converted to acrylonitrile. British Distillers had also worked in this area, converting acrolein and ammonia to acrylonitrile. Now, Idol looked at the research Allied Chemical had carried out but abandoned and thought that other catalysts should be tried. It seemed reasonable to Idol that some of the catalyst systems that Sohio had identified for acrolein and acrylic acid might work.[8] Two or three promising candidates, including bismuth molybdates, were found to work well and the process was quickly optimized, using fluid bed technology. The plant was built by Badger, the contractor with the most experience in designing fluidized bed reaction systems.

The Sohio process shortly became the unique technology for the production of acrylonitrile, with Sohio building plants in the U.S. and licensing the process domestically and globally. The two older acrylonitrile processes almost immediately became uneconomical and plants using those technologies were soon shut down.

Interestingly, this work came at a time when acrylonitrile had just become an interesting monomer, as acrylic fibers and ABS resins had been commercialized not many years before. Sohio had let its scientists work in an area which only became of much greater commercial interest while the research work was going on. There was only a three-year period between Idol's first successful experiment in 1957 and the startup of the first commercial plant in 1960.

Idol then turned his attention to finding other commercial uses for acrylonitrile. This culminated in the development of clear resins from a mixture of butadiene, methyl methacrylate and acrylonitrile. These so-called *Barex* resins found immediate use in beverage bottles for carbonated sodas. However, there was concern about unconverted acrylonitrile in the bottles and they were eventually replaced with a PET (polyester) bottle resin developed by DuPont, which became the standard.

The Sohio acrylonitrile process was extremely profitable for Sohio, generating of the order of 100 million dollars in royalties.[9]

9.4   Ethylene Oligomerization and Shell's SHOP Proess

The search for washing machine and dishwasher detergents had led to the commercial use of 12-carbon lauryl alcohol produced from the fatty acids contained in palm oil and coconut oil from countries like Malaysia, Indonesia and the Philippines. When this alcohol is reacted with ethylene oxide, a very powerful non-ionic detergent is formed.

When ethylene became widely available, chemical companies became interested in synthesizing lauryl alcohol by oligomerizing ethylene to form chains that could

---

[8]Idol (1994).
[9]Ibid.

be converted to alcohols in the lauryl range. Ethyl Corporation first succeeded in producing such oligomers using a Ziegler material, triethylaluminum, reacted with excess ethylene at high pressure to form a range of trioctylaluminum complexes. These were then air oxidized to the alcohol. The result was a range of even-numbered alcohols in a statistical bell curve distribution. The C12 lauryl alcohol is essentially similar to the natural alcohol. The C10 and C14 alcohols also had useful detergent qualities.

Gulf Oil Chemicals, now part of Chevron Phillips Chemical Company, developed a different process to oligomerize ethylene, also using triethylaluminum but in a plug flow reactor to produce a much wider range of even-numbered ethylene oligomers, including detergent-range alpha olefins, higher olefins including heavy materials and tars. Gulf was not interested in detergent alcohols but identified markets for these individual linear olefins, whose distribution and individual quantities could be varied by altering process conditions. Important products were C6 and C8 alpha olefins which found use as co-monomers in linear low density polyethylene. The higher ones are contained in lubricants and in drilling fluids.

Shell started its research on ethylene oligomerization in 1968 when detergents produced from branched, fatty acid-derived natural alcohols did not biodegrade in streams and lakes where waste water containing such detergents was discharged. The initial goal was to produce **linear** alcohols as well as C6 and C8 alpha olefins as co-monomers for polyethylene. The need for a wide range of oligomers looked like an ideal application for a metathesis-type reaction, which would allow the production of a variety of mixtures of alpha olefins. Olefin metathesis is a reaction that involves the redistribution of fragment olefins by the scission and regeneration of carbon-carbon double bonds at high temperatures in the range of 500° Centigrade. A relatively simple reaction, it usually produces few byproducts. Heterogeneous as well as homogeneous catalyst systems can be used. In addition, isomerization takes place, creating internal double bonds, different from Ethyl's and Gulf's processes, where the double bond is at the end of the chain (i.e. alpha olefins). The internal olefins can be reacted with an excess of ethylene and a rhenium oxide catalyst supported on alumina, which causes the internal bond to break, forming a mixture of alpha-olefins with odd and even carbon chain length. The real Shell innovation was in isomerizing the alpha double bonds of the less valuable C4–C10 and C20+ fractions to mixtures of internal olefins in the intermediate range of carbon atom length. These were metathesized to produce detergent range internal olefins.[10]

This more complex process, pioneered and patented by Shell, made it possible to produce a wide number of different products containing various combinations of odd- and even-numbers alpha olefins, some external and some internal. When these materials were hydroformylated (oxo reaction) to alcohols, a range of detergent alcohols could be produced. Since Shell was also an important manufacturer of ethylene oxide, it became a prime source of detergent alcohol ethoxylates. Additionally, in the late 1970s, Shell also became at source of alpha olefins in the C6 and C8 range. The use of metathesis gave Shell an important advantage in being better able to match supply of the different branched chain olefins to market demand (Fig. 9.2).

---

[10]Witcoff et al. (2004).

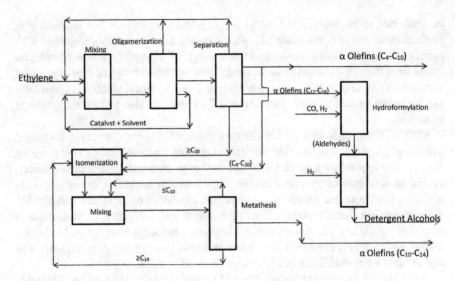

Fig. 9.2 Shell Higher Olefins (SHOP) process with detergent alcohols

During the period when Shell and Ethyl were developing technologies for alcohol-based detergents, other firms such as Monsanto and Exxon were developing another type of detergent intermediate based on kerosene, known as linear alkyl benzene.

In sum, a number of interesting new technologies were developed that competed with natural fatty acids-based lauryl alcohol. As to sources of lower linear alpha olefins, there were Shell and Gulf and later an entirely different source, namely a byproduct of the Fischer-Tropsch plants operated by Sasol with originally German technology in South Africa. Interestingly, these lower alpha olefins had little or no market until polyethylene copolymers entered the market.

Without question, the research conducted by Ethyl, Gulf and Shell led to two entirely new industries, a criterion for the selection of the above technologies.

Metathesis technology was also relied upon in research carried out by Phillips Chemical and Lyondell Chemical during this period, but it was used for a different purpose. Steam crackers produce mixtures of ethylene, propylene and butylenes in typical proportions. The companies operating the crackers generally built them to make ethylene, but would not necessarily need propylene or butylene co-product in the proportions associated with the ethylene produced. A metathesis process would feed propylene to make more ethylene and butylene. Such a process was commercialized by Phillips in 1966. Or it would feed butylenes and ethylene to make more propylene, a variation developed by Arco Chemical, in both cases providing the operator greater flexibility for its downstream operation or sale in the merchant market.

9.5 National Distillers: Vinyl Acetate (VAM)

Vinyl Acetate, a versatile organic intermediate for polyvinyl acetate, polyvinyl butyral and polyvinyl alcohol, had long been produced from acetylene. When ethylene became plentiful and inexpensive, a number of firms commenced research for a process to react acetic acid with ethylene, with ICI in the forefront. The firm developed a liquid phase process using a platinum catalyst and, not being a VAM producer, licensed it to Celanese for commercialization. A plant was built in the U.S. but, soon after startup, major corrosion problems occurred. ICI had apparently conducted corrosion studies to decide materials of construction and had advised use of certain stainless steels. Perhaps Celanese used a different kind, but, in any case, valves, piping and some equipment started to badly corrode. The situation was so serious that it was decided to rebuild some of the affected equipment and piping with titanium. It turned out that the only source of titanium plate was a company called Reactive Metals, owned jointly by U.S. Steel and National Distillers and Chemicals (NDCC). Unfortunately for Celanese, it was informed that every pound of titanium plate was needed for the B_2 bomber program, an interesting fact recently recalled by Joe Pilaro, who worked for NDCC for much of his career. So, the Celanese plant was shut down and scrapped. No other firm apparently attempted to develop a liquid phase process.

National Distillers had, for some time, conducted research to find more uses for ethylene, beyond the polyethylene and synthetic ethyl alcohol it was already producing at Tuscola, Illinois. It turned out that one potentially promising area was vinyl acetate, a co-monomer with polyethylene for making special resins. Accordingly, the firm started to study a vapor phase process operated at modest pressures. Eventually, it was found that a palladium and gold catalyst gave excellent results. Research continued and the firm started to contemplate entering the VAM business.

At that time, there were several small producers of VAM from acetylene, including firms like Borden Chemical and Goodyear. The prospects for a new, very economic process for VAM from ethylene looked very good, with VAM selling for 23 cents per pound versus an estimated manufacturing cost of five or six cents per pound for the new vapor phase process and with the Celanese plant permanently shut down due to lack of titanium for replacing corroded equipment.

Around this time, it was found out that Bayer had built and started up a plant using what looked like similar technology to that of Distillers (Soon to be known as U.S. Industrial Chemicals Corporation) for a vapor phase process. Distillers determined that its patents had priority and sued Bayer, soon winning the case and forcing Bayer to pay a substantial royalty. NDCC built a 100 million pound per year VAM plant, soon expanded to 200 million pounds, and also entered an acetic acid joint venture with DuPont, which also licensed the VAM technology. Distillers also licensed DuPont and Celanese, collecting large royalty payments. Bayer did license some VAM producers, including BP Chemicals, but these firms also had to pay royalties to National Distillers.

Eventually, there were three U.S, producers using the Monsanto acetic acid process and the USI VAM technology: Celanese, BP Chemicals, and USI. (Dupont had by then shut down its plant). Different from many other petrochemical processes, these producers were able to maintain acceptable profitability throughout periods when the cyclical petrochemical industry hit a bad patch. Observers credited this to the uniqueness of the acetic acid and VAM technologies and the fact that the three remaining producers never built expansions far ahead of demand, thus maintaining high operating rates most of the time.

USI's VAM success is eerily similar to SD's success with the Mid Century terephthalic acid process. Both firms carried out exploratory research in an area completely unknown to them. Achieving extraordinary success, both firms created technologies that then dominated the industry. And both had beaten another highly respected firm (Bayer and ICI respectively) by a short time to the patent office.

9.6   Oxychlorination of Ethylene

Technology for the production of vinyl chloride went through several stages. It was originally produced in Germany from acetylene in a simple reaction. This was also the way it was first produced in the U.S. by BFGoodrich and by Union Carbide. When Carbide became an ethylene producer, it reacted that molecule with chloride to produce ethylene dichloride (EDC), which was then cracked to vinyl chloride (VCM), with byproduct hydrogen chloride (HCl) gas, which was hydrolyzed and sold as hydrochloric acid. This production technique was used universally for a number of years with one difference. If acetylene was available, then only part of the vinyl chloride was produced by EDC cracking and the rest by reacting the HCl byproduct gas with acetylene, as before. This was called a "balanced" process.

In the early 1960s, several companies looked at another option, which would use the HCl rather than having to sell it at a low price or even dispose of it via neutralization, which was even more expensive. One alternative was to oxidize the HCl back to chlorine using the so-called Deacon process that employed a cupric chloride catalyst, but that was found to be uneconomical. Then, some of the companies thought of using the Deacon catalyst in the presence of ethylene and oxygen and that looked like a winner. This reaction is highly exothermic and so a means had to be found to run the reaction. Dow and Stauffer decided to use a series of fixed bed reactors to carry out the process in a staged manner.

Goodrich teamed up with Badger, known for its expertise with fluid beds, to design a process using that technology. It built the first plant, a 400 million pound per year facility at Calvert City, Ky. Several other companies developed their own fixed bed oxychlorination technology, notably Ethyl Corporation, Stauffer Chemical and Toyo Chemical in Japan[11] (Fig. 9.3).

The importance of oxychlorination was that there are few places where acetylene is available and companies don't make acetylene on purpose any more (except for welding). So, new VCM plants had to dispose of the HCl, which, as mentioned

---

[11]Peter H. Spitz op.cit. 403–404.

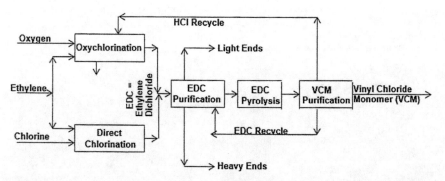

Fig. 9.3 Vinyl chloride production via oxychlorination

above, is often a wasteful proposition. Oxychlorination, using additional ethylene was obviously the answer (Except in China, where acetylene from coal is still used in a balanced operation.).

The oxychlorination process was therefore widely practiced and available from several licensors to global vinyl chloride producers. It changed the global vinyl chloride industry and therefore deserves to be among the technologies selected for this chapter.

Some years later, ICI was developing a process to produce vinyl chloride directly from ethane. One or more plants were built by a successor company to ICI but this technology never gained wide acceptance, most likely because its economics were not attractive enough to replace the route using oxychlorination.

References

Green, Mark M. and Harold A. Witcoff (2003) *Organic Chemistry Principles and Industrial Practice* Wiley-VCH GmbH & Company. 244.

James D. Idol (1994) *Oral History*. Chemical Heritage Foundation. Philadelphia, Pa.

James F. Roth (1995) *Oral History*. Chemical Heritage Foundation. Philadelphia, Pa.

Peter H. Spitz (1988) *Petrochemicals. The rise of an industry*. John Wiley & Sons. New York.

Vladimir Haensel. (2006) *Biographical Memoir*. Nat. Acad. Sciences Volume 88.

Weissermell K. and Arpe H.J. (1978) *Industrial Organic Chemistry* Verlag Chemie Weinheim-New York 157.

Witcoff, Harold A., Bryan G. Reuben, Jeffrey S. Plotkin. (2004) *Industrial Organic Chemicals*. John Wiley & Sons, New York. 116–117.

Chapter 10
The Leading Research and Engineering Firm in Petrochemicals

Abstract By the mid-1970s, SD had become the leading firm designing petrochemical plants. It now had a catalyst manufacturing arm and an ability to carry out complete engineering and procurement functions and to construct some of the plants it was designing. In addition to designing plants based on licensed SD technologies, the company acquired from other firms the rights to offer a broad range of additional technologies. A number of examples of successful projects are presented. A book edited by Ralph Landau became a manual for chemical engineers working in the industry.

By the end of the 1960s, petrochemicals had largely replaced the organic chemicals and polymers previously made from coal, alcohol and wood and had become the perfect material for a large number of products (e.g. polystyrene foam, agricultural films, tires from synthetic rubber, etc.) that were responsible for the double digit growth rate of polymers such as polyethylene, styrene and polyamides. It is fair to say that the transition to a new industry had, by that time, been largely completed. This was, however, not the end of intensive research on petrochemical technology. The entry of so many firms as participants in this new industry led to increasingly intense competition as companies sought to gain manufacturing cost advantages. This could, to some extent, be achieved through scale, but more advantageously through yield improvements and other modifications in the production technologies, e.g. development of higher selectivity catalysts. Breakthrough technologies were the best means, with some of these described in the two previous chapters. Then, by the late 1970s, the petrochemical industry had started to have reached a level of maturity.

Scientific Design Company had now reached maturity as a research and engineering firm. It had moved its research activities to Little Ferry, New Jersey and had built up a relatively large engineering staff at 2 Park Avenue. Its French subsidiary, Société Française des Services Techniques, had a solid sales and engineering capability, SDPlants, Ltd. had been set up in London, and a Japanese office had also been established.

In every respect, Scientific Design now had the same capabilities with respect to process design, engineering design, plant layout, equipment purchasing, construction and startup assistance as such venerable companies as M.W. Kellogg, Foster-Wheeler, C.F. Braun, Lummus, Fluor and others. Of course, the entirety of SD's

© Springer Nature Switzerland AG 2019
P. H. Spitz, *Primed for Success: The Story of Scientific Design Company*,
https://doi.org/10.1007/978-3-030-12314-7_10

Fig. 10.1 Brass Plate taken from SD's entrance door at 2 Park Avenue, New York

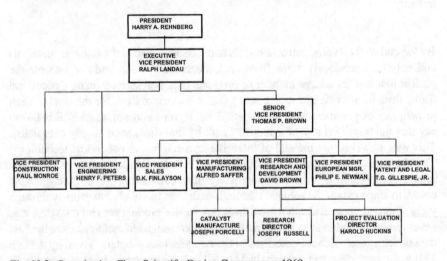

Fig. 10.2 Organization Chart Scientific Design Company ca. 1960

experience was in chemicals, while its competitors were primarily active in the petroleum refining, natural gas, fertilizer, and other heavy industries. This was a heady accomplishment, carried out over a period of roughly twenty years and the company's logo became well known (Fig. 10.1).

A strong sales team headed up by Landau, Rehnberg and Tom Brown, who died unexpectedly, had brought in hundreds of projects and its legal team already had a strong record in developing and defending patent positions (Fig. 10.2).

Fig. 10.3 Massachusetts Institute of Technology: Main building and courtyard. *Source* MIT

As two of SD's founders had received Ph.D.'s in Chemical Engineering from MIT, the firm was very partial to hiring MIT Chem. E. (Course X) graduates, many of whom had advanced degrees (Fig. 10.3).

Ralph Landau ScD 1941	Robert Egbert ScD 1941	David Brown SCM 1940
John Colton BS 1948	Peter Spitz SCM 1949	George Marlowe ScD 1950
Joseph Russell ScD 1955	Theodore Stein ScD 1955	Robert Davis ScD 1955
David Koch ScM 1963	William Long SCM 1960	John Lutz ScD 1943
Ernest Korchak ScM 1961	John Schmidt ScD 1963	Jon Valbert (Asst. Prof.)
Manfred Gans ScM		

Several of the SD engineers interviewed for this book remembered David Koch well, even though he was not at the firm very long. One of them shared a room with David Koch and said that he usually put his paychecks in his desk and seldom remembered to cash them. David and his brother Charles currently head up Koch Industries, which is one of the largest private companies in the US, with major activities in oil and petrochemicals.

SD hired a large number of bright chemical engineering graduates from other universities with strong chemical engineering departments. These included, among others, the University of Michigan, Princeton, Cornell, University of Wisconsin, and the University of California-Berkely, SD had established itself as a developer of unique technologies, including the Mid-Century process, the cyclohexane oxidation technology and isoprene, and it had also developed a position in several technologies where SD had been able to offer clients licenses and engineering services to build petrochemical plants when they could not obtain licenses for these technologies from established producers. (See Chap. 6) A fairly robust stream of royalties had been established, though most of SD's income had to come from engineering fees to support its large staff. By the early 1970s, SD's annual income was in the range of $70–80 million a year, as claimed in an SD brochure published at that time.

Two of the founders, Landau and Egbert, were not only accomplished chemical engineers, but had also been very inventive, as shown by the impressive list of patents issued in their names over the first three decades of SD's existence. In addition to his entrepreneurial talents, Landau also kept extremely close to the research work, making sure that the work was done efficiently and that costs were kept low, particularly in the early days of the company. Landau, Dave Brown and Al Saffer formed the committee that decided on the most promising areas of research that should be funded and they particularly concentrated on the patentable aspects of the inventions. John Colton, considered one of the most brilliant chemical engineers at SD, was often consulted on the likelihood and viability of a promising laboratory result becoming a commercial reality. Joe Russell and John Kollar developed a strong record in originating new ideas and then guiding the research work. Fred Gans became a key member of the firm with responsibility for startups and guarantee runs.

Most if not all of the staff were in awe of Ralph Landau, though few knew him well. Neil Yeoman in his memoirs of SD wrote, "Landau's genius was a critical factor in (SD's) rise and fall. The operation was a team effort, and there were many outstanding participants, but the unique achievement could not have happened, had not the ultimate authority for all critical matters been a man of such extraordinary talent." Landau was known to be egotistical and overbearing to the point that Ron Cascone, a respected SD engineer, recently recalled that some people thought Landau had a "Napoleon complex". Considering what that historic general accomplished, that was perhaps a backhanded compliment.

10.1   The Diverse Nature of the Projects and the Exciting Atmosphere that Prevailed

The author worked at SD from 1956 to 1964 and can well remember a real *esprit de corps* and animation among the researchers and engineers who worked at SD. Landau remarked on this in his book published in 1994:

> Joel Kirman, who joined Scientific Design in 1962, captured that sense of excitement in a speech in which he described the company as a "veritable anthill of activity." The laboratory and pilot plant would be working on ten or more processes. There would be continuous interaction with the process engineers and sales people to determine the best routes to redirect investigations. Judgments were constantly required on where to accelerate activity and where to reduce it; at what sales or managerial level to notify industry of either pending or favorable results; how to deal with the discovery of weaknesses in any process as it then stood, how readily and with what factors of safety could scale-ups be performed. And so forth.[1]

---

[1] Landau (1994)

Brian Ozero, the son of Ukranian immigrants who had fled to Canada after the Russian revolution, was educated in Canada as a chemical engineer. After graduation, he worked for Shell Chemical in England and then returned to Canada. When he applied to Shell Canada for a job, his interviewer allegedly said "we will be glad to hire you, but you should really try and join Scientific Design because that's where all the action is right now".

Success in its research and engineering activities still required constantly looking for projects to keep its large staff busy. A substantial amount of engineering work came from ethylene oxide and from maleic anhydride licensees. PVC and vinyl chloride, chlorinated solvents and cyclohexane oxidation also provided engineering work. But SD decided that with its large staff it needed to look for additional work by making its engineering and R&D capabilities available to others, similar to what contractors like MW Kellogg. Fluor, Badger and several others were doing. And because SD became well known as an engineering firm specializing in petrochemicals, this gave it an advantage in securing projects from chemical companies that needed SD's well-honed chemical engineering capabilities to realize new projects. A number of such projects illustrate the breadth of SD's successful efforts in securing such assignments which, in a number of cases, then gave SD opportunities to offer third party technologies to other clients.

- In 1963, SD designed and constructed a synthetic rubber complex for Goodyear Tire and Rubber Company in Beaumont, Texas. It included plants for the production of isoprene, polybutadiene and polyisoprene.
- In 1963 SD designed and constructed a vinyl acetate plant for Monsanto in Texas City, Texas. The process had been licensed by Monsanto from a third party, with SD making engineering improvements (see photo).
- In 1962, SD carried out the engineering, procurement and construction of a synthetic glycerin plant for Olin Mathieson in Brandenburg, Kentucky. This was a first-of-a-kind plant based on a third party's knowhow.
- In 1966, SD designed, engineered and constructed a high pressure polyethylene plant for Dow-Unquinesa in Tarragona, Spain.
- In 1962, SD designed, engineered and constructed a 22 million pound per year adipic acid plant for Rohm and Haas Company in Louisville, Kentucky. This plant was based on the conventional nitric acid oxidation of cyclohexane.
- In 1962, SD carried out process and development work for Merck and Company on a plant to make thiabenzadole, a proprietary pharmaceutical.
- In 1965, SD carried our process design, engineering and procurement for a nylon intermediates plant for the Chemstrand fiber subsidiary of Monsanto.
- In the mid-1960s, SD designed, constructed and successfully started up an ethylbenzene-styrene plant for Société Nationale des Petrôles d'Acquitaine.
- In the early 1960s, SD designed plants for the production of ethylene from ethanol for Solvay & Company in Brazil and for Manzoor Chemical Company in Pakistan. These plants used a process developed by the Air Reduction Company.

- In 1964, SD designed a plant for the production of ethyl and butyl glycol ethers for ICIANZ in Sydney, Australia. The process was developed by SD using information in the Public Domain.
- In the late 1960s, SD designed and assisted in the construction of ethylene dichloride plants for Mitsubishi Chemical, Chinese Petroleum Company and Pemex. These plants used knowhow provided by Monsanto.
- In 1968 SD developed a process to make aniline from phenol and designed a 44 million pound per year plant for Mitsui Petrochemical Company.
- In 1964 SD designed a plant to make cyclohexane from benzene, using internally developed technology.
- In 1967 SD designed and started up an ethylbenzene plant for Yawata Chemical Company in Japan.

Four interesting examples of the diverse opportunities that came SD's way are described below, naming the engineers who worked on the projects:

- Marshall Frank, who had joined SD directly after graduating as a chemical engineer from Cornell, recently related that in the late 1960s, SD was asked to design a propylene oxide plant based on the soon to be obsolete chlorhydrin process for a firm in England. It accepted the project and used B.I.O.S. report No. 1059 describing the German technology for developing the design. This was at a time when SD was developing its patented propylene oxide process described in Chap. 7, but SD's engineering department was undoubtedly pleased to have a project to keep its staff busy.
- SD was usually more than willing to undertake a project even when the company had no prior experience in the area, yet believed it could meet the challenge. As Joel Kirman recently recalled, that was the case in the early 1960s when W.R. Grace approached SD and asked whether it could design a vinyl chloride plant for Grace in Peru. Grace wanted to establish a chemical industry in the country where it already had a large sugar plantation. With a large local alcohol production plant based on sugar fermentation, Grace saw that it had a useful raw material. The company's engineers obtained a government B.I.O.S. report that showed how the Germans produced ethylene from ethyl alcohol. The Grace plant was at the Pacific Ocean and it was found that the brine was suitable for Chlor-alkali production. German technology could then be used to make ethylene dichloride (EDC). SD was asked whether it could design an EDC cracker to produce vinyl chloride (VCM). This is a pretty straight forward process and so SD found a furnace manufacturer who had the appropriate experience. And so Grace built a VCM plant as well as a PVC plant as part of the *project*. It is not known whether SD designed the PVC plant.
- Monsanto was developing an ethylene oxychlorination process, based on oxygen or air and using a fluidized bed. It was not ready for commercialization and asked SD to review the technology and finalize the flowsheet and equipment selections. Brian Ozero was in charge of this work and presented a paper on the process at the March 1971 AIChE conference in Houston. One or two plants were built in Japan and Mexico, but the process was not a big commercial success.

- Neil Yeoman related the following. The Swedish company Mo Och Domsjo, which had built an ethylene oxide plant with SD technology, approached SD and asked whether it could help in commercializing an ethylene diamine process which the Swedish firm had studied in its laboratory. This is a rather small-scale chemical with applications for various syntheses, including pharmaceuticals and chelating agents. SD accepted and used its background in piloting and scale-up to design a small commercial unit. The plant was soon built to European standards with very small piping and packed columns. For some reason, the firm used a different packing than SD had specified and had oversized the columns. The plant did not work well and SD engineers were asked to fix the problem. When the originally specified packing was procured and substituted, the plant met all expectations and could operate at well over design capacity.
- SD had other projects in pharmaceuticals. A plant for a confidential chemical was designed and built for Merck and Company in Rahway, N.J. in 1962.
- By the late 1960s, SD had accumulated a large number of technologies from third parties which could then be available from SD, who would perform engineering and such other services as might be required (Source: Scientific Design Brochure).

10.2   Third Party Processes Available Through Scientific Design

Isobutyl alcohol	polychloroprene	xylene isomerization
n-butyl alcohol	polyols	p-xylene crystallization
sec-butyl alcohol	propylene oxide	terephthalic acid
chloroprene	toluene diisocyanate	isophthalic acid
ethanol	trichloroethylene	polyethylene
ethylene diamine	polyisoprene	vinyl chloride
2-ethyl hexanol	vinyl acetate	bisphenol-A
hexamethylene diamine	glycerine	isodecanol
methyl ethyl ketone	benzoic acid	isooctanol
malic acid	oxo alcohols	isopropanol
methyl chloroform	cumene	polystyrene

10.3   Scientific Design Diversified into Coal-Related Work as the Oil Shocks Impact the US Energy Situation

The two oil shocks in the 1970s changed the US energy picture dramatically. As crude oil prices first rose from $3 per barrel to over $30 and then much higher, electric power plants switched from oil to natural gas or coal, greatly increasing gas demand and price. Meanwhile, gas production was lagging, with relatively few new discoveries. As a result, the supply-demand balance for natural gas totally changed, with gas becoming scarce in some regions of the country and some industrial users cut off from their gas supply. Eastman Chemical, located in Kentucky/Tennessee had to shut down some operations and decided on gasified coal as a chemical feedstock. Coal was suddenly looking like a very desirable raw material and the US had a huge supply, both in Appalachia and in some Western states.

This was a boon for coal companies and a wake-up call for industry and for Congress. The Carter administration was understandably very concerned about the rising price of the two main sources of US energy (crude oil and natural gas) and loosened the purse strings for research and development on the country's largest resource: coal. The Department of Energy (DOE) was mandated to initiate projects for the utilization of coal in places where either oil or natural gas were being traditionally employed. The government thus challenged engineering firms to work on technologies that could employ coal for the production of fuel gas, synthetic crude oil and chemicals. And at the state level, Congressional representatives from states with large coal deposits secured funding for projects as soon as they could find them, sometimes justifying essentially duplicate projects by stating that coals were different from state to state, with Kentucky coal completely different from Powder River coal in Wyoming.

SD was quick to join the queue for projects receiving government funding. It would, of course, have to compete with a number of other engineering firms, but SD had a tradition for research which many other engineering firms lacked. And so, SD's sales and engineering people became frequent users of the Amtrak train to Washington, calling on DOE.

Ironically, it was only thirty years earlier that Germany was using coal and lignite as the main source of fuel and raw material for organic chemicals production. It was time to dust off the reports on Germany's use of coal, as well as spending DOE money to looking at ways to use coal instead of oil or gas.

The most effective technology to create a high Btu fuel gas is the gasification of coal to make so-called synthesis gas, consisting of a mixture of carbon monoxide and hydrogen, with carbon dioxide as a byproduct. With further processing in two steps, the synthesis gas is converted to methane, then called "synthetic natural gas" (SNG). This could be directly used in places where natural gas supplies had been cut off to consumers. (For example to those with so-called "interruptible" contracts at low prices, where the supplier had the right to withhold supplies in case of gas supply shortages.) SD developed its own process for the production of SNG.

There was also a significant effort by DOE to sponsor technology that would produce a "synthetic crude oil" from coal. The main technologies being developed were processes that would use a hydrocarbon solvent to effectively "liquefy" coal, known as SRC-1 and SRC-2. This type of process was different from German technology and was treated to substantial amounts of research money. Looking back, it is fair to say that nothing ever came from this work.

Fred Gans, Marshall Frank and Ron Cascone, with others, became the SD group that travelled to Washington to compete with other engineering firms for DOE-sponsored work. Some of the projects won by SD in the mid-to late 1970s were:

- A technical and economic evaluation of a Battelle technology for making high BTU gas in a pilot plant using the Agglomerated Ash process.
- A similar project for the so-called Hygas process under development by the Institute of Gas Technology in Chicago, Illinois using a stage fluidized bed steam gasifier, which would maximize methane production.
- A process that would gasify coal or biomass to make medium Btu fuel gas.
- Review of a test program covering a Texaco coal gasifier and hydrogenation plant for vacuum bottoms from a 200 barrels per day SRC demonstration plant.
- Review of a Ft. Lewis, Washington pilot plant operating a SRC process on Kentucky and Pittsburg seam coals.
- Evaluation of process design for a grassroots SRC-II process commercial plant for Pittsburgh & Midway Coal Mining company.
- Audited technology for Allis-Chalmers rotating kiln process for low BTU gas from coal for combined cycle operations.

The fact that SD was selected for these and other projects of this kind speaks well of the qualifications it brought to these competitive projects. Other competitors often did not have experience with pilot plant design and operation and with the chemical engineering "unit operations" inherent in these processes.

10.4 Scientific Design "Wrote the Book"

Landau believed that he and his partners and co-workers had, over the years, created a highly organized method of going about the business of developing new technologies, designing new petrochemical and other plants, carrying out the purchasing and construction, assisting in plant startup and helping to improve and optimize plant operations. No other engineering firm could come close, as far as experience in petrochemical plant design was concerned. Landau asked a number of his co-workers to write chapters covering the various aspects of carrying out the various tasks required and then published a book entitled *The Chemical Plant. From Process Selection to Operating Plant*,[2] incorporating these chapters. The contents included:

---

[2]Landau 1966

1. Introduction: Ralph Landau
2. Process Development and Commercialization: David Brown
3. Process Evaluation: Peter Spitz
4. Process Licensing: Gerson S. Schaffel
5. Selection of Contractor Scope and Contract Types: D. Kenneth Finlayson
6. Plant Layout and Site Considerations: Robert Merims
7. Process Design: Harold A. Huckins and Theodore W. Stein
8. Project Engineering and Management: Russell G. Hill
9. Engineering and Purchasing: George J. Marlowe
10. Field Construction: Harry Peters
11. Cost Control and Estimation: John H. Lutz
12. Plant Start-Up: Manfred Gans and Frank A. Fitzgerald
13. Successful Operation: Improvement and Application to Future Designs: Gregory F. Vinci.

The Introduction to the book points out that the engineering of chemical plants is considerably more complex than that for petroleum refinery projects, though the latter were originally responsible for the development of modern chemical engineering practice". Chemical plants usually make single products at high yields and purities, require complex catalysts, must operate under wider extremes of temperatures and pressures and may require greater flexibility to meet competition. They often have solids handing problems, face greater corrosion and toxicity problems, and handle a much more complex set of raw materials, intermediates, solvents, and catalyst systems.

The book is well illustrated, with examples of flowsheets, equipment sketches, production cost estimates, organizational tables, construction schedules, startup schedules, and photographs. At the time, it was considered an important reference for chemical engineers considering a career in the process industries or working on a specific project.

10.5   Linking Up with MIT

Around this time, MIT asked SD Management whether it would be willing and interested in having MIT establish a station of its Chemical Engineering Practice School at Scientific Design Company. This would be the first time when such a station would be at an engineering firm rather than in the plant of an operating company. The purpose of moving Masters' degree candidates through an SD station would be to acquaint them with contemporary research and engineering practices, including process and project evaluations, process design and equipment selection problems. Since SD had always had a close relationship with MIT, having hired many of its advanced degree graduates, it immediately agreed and the station was set up and run for a number of years thereafter.

10.6   The Stage Was Set

Landau and Rehnberg at this point believed that they had succeeded beyond their original expectations, managing an engineering firm that had become well known throughout the global industry and credited for its inventiveness and successful risk-taking. Its only competitors in the realm of engineering firms with strong research arms were UOP and the French Petroleum Institute (IFP), but these firms were more oriented toward refining technology than petrochemicals. The engineering contractors building refineries, ethylene, ammonia, and methanol plants were, of course, much larger in terms of employees and revenues than Scientific Design, but these firms carried out fewer chemical projects.

The only goal not yet achieved, but always at the back of Landau's mind was to get into actual production, i.e. to become an operating company. He knew that that was the only way to build value into the firm so as to have a steady stream of income not dependent on engineering income or royalties. He saw in the epoxidation technology a way of finally getting there. Already In the mid-1960s, he and Rehnberg had decided to form a firm they called Halcon International, which was to be a holding company that would have SD as an engineering subsidiary, SD research as another subsidiary, and Halcon Chemical as a (future) manufacturing arm.

Engineers and researchers working at SD gained no knowledge of the firm's income or profitability. Clearly, the firm's policy was that engineering was a profit center and could not be supported by royalty income. But there was always a great deal of secrecy where money matters were concerned. Ernie Korchak recently remembered that when he was asked to develop a business plan, he was unable to receive information about the P and L of historical projects and asked rhetorically, how can you do a business plan if you can only develop the cost side of the equation?" But a privately held, entrepreneurial firm had to be closely managed and its capital carefully husbanded. In this, the founders succeeded.

References

Ralph Landau. (1994) *Uncaging Animal Spirits.* The MIT Press. Introduction.
Ralph Landau (1966) *The Chemical Plant. From Process Selection to Operating Plant.* Reinhold Publishing Corporation. New York.

Chapter 11
Oxirane and the Creation of Halcon International

Abstract With the successful propylene oxide technology, Landau had finally found a vehicle to get into the business of manufacturing chemicals. A new organization is created, called Halcon International, with divisions for manufacturing, R&D, engineering and catalyst manufacturing. A partnership called Oxirane Corporation is formed, with Arco Chemical to jointly commercialize the new propylene oxide process which Arco had also been working on. Several plants are built in the U.S. and in three foreign countries. The venture becomes a resounding commercial success, with Halcon receiving very substantial annual income. Several experienced chemical company executives join the firm, as its ambitions had grown.

By 1963, Scientific Design Company was well established and the owners decided it was time to reorganize the operation. A new company, Halcon International, was founded, which would be the holding company for three subsidiaries:

Scientific Design Company: the engineering and construction arm. Headed up by Paul Monroe, this division was by the 1960s able to offer complete lump sum services (process design, engineering, procurement, construction and startup assistance) for projects in the U.S., Canada and Europe. Startups were generally carried out by a group headed by Manfred Gans, a chemical engineer who was born in Europe and had escaped the Holocaust when his family fled to the United States.

Halcon Research & Development Corporation: A research center would be constructed in Montvale, N.J., which could ultimately house 200 people. The company's research staff was very strong, with Joe Russell and John Kollar having starring roles. Ernie Korchak, who had joined SD in the 1960s from ICIANZ in Australia, took over as President of Halcon R&D. Kollar had been the lead researcher for both cyclohexane oxidation and epoxidation and was held in the highest esteem. John Schmidt, to whom Kollar reported, described him as a sort of large rambling "genius," looking like an NFL linebacker and very sure of himself. When making an invention, he always wrote a long report describing exactly what it was about and why it was important.

Catalyst Manufacturing Corporation, headed by Al Saffer, was a separate entity.

Halcon Computer Technologies: SD believed it was one of the first engineering companies to use an IBM 360/40 computer system for process design and flow-

© Springer Nature Switzerland AG 2019
P. H. Spitz, *Primed for Success: The Story of Scientific Design Company*,
https://doi.org/10.1007/978-3-030-12314-7_11

sheeting, engineering design and commercial uses, such as project management and control and cost estimates. Drawings were generated, representing an actual engineering and purchasing schedule giving a flow plan of all activities and events that must be accomplished on an actual project.

When it was likely that the company would become a manufacturer of chemicals, a fourth subsidiary was added:

Halcon Chemical Company: Manufacturing.

The staff coming into this organizational restructuring could, of course, move back and forth between these entities.

In 1959 or 1960, Landau and Rehnberg had let a number of key people participate in the success of the operation. Then, in 1969, seven top management people were able to purchase 13,684 shares at $48 per share, with a small down payment and the rest financed at 4%. Joe Russell, head of research, bought the same number of shares at the same price somewhat later. Marc Millard, a senior partner at Loeb Rhoades and Company, an eminent Wall Street firm, in 1969 valued the firm at $90 million, based on capitalizing research costs and applying a figure of 40 times earnings! Landau and Rehnberg also made available 3000–5000 shares (each) to a number of other employees, apparently in the form of so-called "phantom stock", which was not actually bought but gave recipients the right to sell at an appropriate time.

Scientific Design in the mid-1960s was clearly a major success, having established a highly creative research organization and an extremely competent engineering department. The Mid-Century process and the cyclohexane oxidation, and epoxidation technologies owed their success to management having hired and given strong support to creative researchers and to having hired a number of chemical engineers who used the new techniques of their profession to turn laboratory findings into viable process technologies. What was happening at Scientific Design was becoming well known and talked about in the industry.

But now Landau and Rehnberg started to think seriously and confidently about the future and getting into chemical manufacturing. Their thinking was well expressed in Landau's speech to the Newcomen society in 1978.

> Why were we so interested in chemical manufacturing when we had a good licensing and engineering business going? The reason for this desire on our part were complex but included: a. recognition that royalty income alone could not pay for the increasingly greater costs of research and development for the new technology; b. recognition that pure service organizations would never have great capital value; and c. realization that the return on really creative new technology would be greater by participation in the manufacture of the products under the exclusive protection features of the free world's patent system, rather than by nonexclusive licensing, (often) for modest royalty to many smaller plants often uneconomical in size[1] (Fig. 11.1).

In retrospect, it seems obvious that Landau could now feel that the only logical next step would be to get into manufacturing and become an operating company. The idea of "going public" was never far from his mind, but the company had not been ready for that. The owners knew that engineering firms would never have the kind

---

[1]Landau (1978).

Fig. 11.1 Ralph Landau at the Newcomen Society event *Source* Newcomen Society in North America

of valuation, in terms of price-to-earnings (P/E) ratio and earnings expectations, that would accrue to a company with steady manufacturing earnings. Professional firms are only as valuable as their employees and, as the saying goes for this type of firm, "the company's assets go down the elevator every night". So, if the owners wanted to "cash out" at some point, the company would first have to develop a large and steady source of earnings and then go public. And that was the eventual goal.

Landau was familiar with the history of the company founded by Dr. Haldor Topsoe, a Danish scientist, who had been active since before the war in developing chemical catalysts for different non-organic processes, including various hydrogenations as well as for the manufacture of ammonia. A large percentage of worldwide ammonia plants were using Haldor Topsoe catalysts and many were licensing the Haldor Topsoe process. The company had gone public in 1972 and, somewhat later, had an enviable net income around 500 million Danish Crowns (50 million dollars) per year. Interestingly, Dr. Topsoe said in an interview in 1999 that he had at one point wished that he had also carried out research in the areas of Landau's company, but did not have the resources to do so![2] This obviously speaks volumes about SD's earlier accomplishments, achieved with essentially no capital resources.

11.1 The Creation of Oxirane

When SD's research on propylene oxide became successful, its management knew that it had a winner on its hands. Could this finally be the way that SD could become a manufacturing company? Landau had tried back in the early fifties with the direct ethylene oxide process, but it was too early and his attempts had quickly failed. Then, with the Mid-Century process, there was another attempt, but the need for cash and the attractive offer from Amoco Chemical again precluded SD from becoming an operating company. Now, another opportunity presented itself. Of course, SD would need a partner to fund the development of the new epoxidation technology.

In the course of talking to potential partners, SD's patent attorneys came across a disturbing development. Arco Chemical evidently had been doing research in the same area, though only on the C_4 version of the technology. There was considerable consternation, immediately replaced by the conviction that the two companies should merge their interest and form a partnership to exploit all aspects of the novel hydroperoxide-based epoxidation technology. SD was convinced that it had the dominant patent position, which should make it easier to convince Arco to get into a joint venture. And this would finally let SD become a partner in an operating company, provided that Arco would not only put up the capital for its share, but also help with financing SD's share. Evidently, an interesting negotiation lay ahead. Arco Chemical at that time was a very small subsidiary of Atlantic Richfield Corporation, a large oil company with partial ownership of Alaska's North Slope oil. Contrary to the chemical divisions of most other large oil companies, Arco Chemical had little experience in the petrochemicals industry, though it evidently had a good research arm. Harold Sorgenti recalled that his group had developed a linear alkyl benzene process and a toluene dealkylation process before starting its work on propylene oxide.[3] It would be interesting to compare notes with Arco's researchers, SD researchers told each other.

Landau looked at the management structure of Atlantic Richfield and Arco Chemical, looking for an approach and then he smiled. Arco Chemical's president was Robert Bent, a classmate of Landau's at Penn before the war. It was time to place a telephone call.

Discussions between SD and Arco Chemical started off quickly. The new and inexperienced chemical firm was quite eager to consider a partnership. A concern about a recent failure of its Sinclair Oil subsidiary with an ammonia plant using new technology made Arco nervous about commercializing a new process, a state-of-mind recently described to me by Ed Zenzola, a retired Arco Chemical executive. However, both Atlantic Richfield's chairman

---

[2]Topsoe (1999).
[3]Sorgenti (2003).

Robert O. Anderson and Robert D. Bent had an entrepreneurial nature which facilitated the concept of partnering with an unconventional technology-oriented company such as Scientific Design. Interestingly, Harold Sorgenti, who had just been promoted from Arco Research to a top management position under Bent, could not understand the logic of combining the two technologies, since for Arco the PO/TBa version, which they had researched and patented, was the one that perfectly fit their business, as the co-product alcohol could be blended into Atlantic Richfield's gasoline as a high octane anti-smog additive.[4] However, both Bent and Anderson were persuaded by Landau to form the partnership and that is what happened. The two parties took six months to come to a complete agreement.[5]

In 1966 the partners set up a new company called the Oxirane Corporation with headquarters in Princeton, N.J. to manage the business of the new technology, including the R&D, engineering and marketing. The name "Oxirane" was appropriate as it covered both propylene oxide and ethylene oxide (the earlier SD successful technology) which were, in chemical terms, members of the *oxirane* family. A manufacturing partnership was set up in Bayport, Texas to plan the construction of the first plant.

The financial arrangement negotiated between Halcon and Arco Chemical was extremely favorable to Halcon, which was in no position to put up or borrow from banks the capital required to make it a 50% partner in the venture. Arco and Bob Bent were, however, keen to proceed with the venture and were persuaded by Landau, Dave Brown and Joe Russell that the technology would lead to successful commercialization. The financing was set as follows:

– Arco provided 90% of the capital and Halcon 10%.
– Arco's 90% was in the form of 50% cash and 40% a non-recourse loan to Halcon.
– The Arco loan to Halcon would be repaid with a series of notes with 10% interest and repayment over the next several years, using the cash flow Halcon received from the venture.

Based on projections, Arco Chemical's rate of return on the first plant was estimated to be of the order of twenty percent.[6]

11.2   Construction of the Bayport Plant

The two partners were eager to get started. The first question was where to build the first plant and that was easy. It should be in the United States and adjacent to an Atlantic Richfield refinery. So, it was quickly decided that the plant should be built at Bayport. The next question was which epoxidation route should be used. That was fairly easy, as well. If Oxirane were to use the PO/Styrene co-product process, the new company would have to become a styrene producer and marketer,

---

[4]Ibid.
[5]Ralph Landau. Op.cit.
[6]Landau (1994).

competing with a number of existing companies making this product. With a co-product process, the total cost of production has to be allocated to the two products, in that case propylene oxide and styrene. If the existing PO producers were to lower their market price to meet the entry of a new produce, Oxirane would have to meet that price, shifting more of the production cost to styrene. This could make market entry in styrene somewhat difficult, particularly for a first of a kind plant. But if the tertiary butyl alcohol/PO co-product route were used, the alcohol could be blended directly into Atlantic Richfield's gasoline. That worked well, because legislation had recently required refiners to blend oxygenates into gasoline to meet EPA standards for reducing smog emission from automobile exhausts and vaporized solvents. This would make the PO marketing situation much easier. Finally, what size plant? It was decided to design the plant to make 160 million pounds per year of propylene oxide, expandable later if successful. This first-of-a-kind plant would be orders of magnitude larger than the first terephthalic acid or nylon intermediates plant, but the partners apparently had a lot of confidence—particularly Arco, which was putting up essentially all of the money.

The process was based mostly on Arco's research, with some input by Halcon. Both partners collaborated on the plant design, though most of the detailed process engineering was carried out by SD engineers Brian Ozero and Neil Yeoman under the direction of Dick Kaplan. The flowsheet is fairly complex, with a large number of unit operations. Much time was spent considering the complicated heat balance, with the result that a number of opportunities were found to generate both low and high pressure steam to reduce total utility consumption. As it was a new process, the design was fairly conservative to allow for unforeseen problems. An estimate for the total cost of the plant had been made fairly early and had been used to get the financing established. When the design was complete and the plant cost was re-estimated, the cost was $100 million or so over the original estimate. Part of the reason was the decision to use quite expensive construction materials in some sections of the plant. The designers were told that the plant would never be approved at that cost. So, as Brian Ozero recently recollected, he and Neil stripped out as many equipment items as possible (e.g. duplicate pumps, bypass valves, extra tanks, etc.) to get the cost down. Finally, they received the go-ahead and the plant was constructed and started up with little or no problems! In fact, the plant came up to design capacity relatively quickly and was operated continuously for 69 days, reaching 120% of its design capacity. Oxirane was in business (Fig. 11.2).

Shortly after startup, it was decided to debottleneck the plant to 190 million pounds and to build a second plant of the same kind sized at 300 million pounds, which started up in 1971. There was actually a small explosion at the first Bayport plant, but that did not deter the partners, according to a recent discussion with Morris Gelb, a high level executive at Arco Chemical/Lyondell. That plant was duplicated and the second plant started up in 1974.

With demand for oxygenates soaring, the operation was very profitable. This effectively gave Oxirane substantial price control and resulted in other domestic PO producers other than Dow to quit the industry over the next few years!

Fig. 11.2 Bayport propylene oxide plant under construction. *Source* Chemical Week

This is a very interesting point. When developing a brand new process requiring very substantial investment in research and development and in the construction of a first-of-a kind plant, the question must be asked: Will the new process have "shutdown economics?", i.e. have a cash cost of production so low as to force the existing producers to shut down their presumably now uncompetitive plants? If so, the investment for developing and commercializing the new technology will pay off handsomely. This happened in several cases in the petrochemical industry (Examples: UOP Platforming, SoHo acrylonitrile, Monsanto acetic acid). Evidently, this was the case for the new epoxidation technology, which was responsible for the shutdown of almost all existing chlorohydrin-based propylene oxide plants. But many times a new technology will not be commercialized because its economics are just not good enough to obsolesce existing producers operating at their cash cost of production.

As soon as the technology was proved out at Bayport, planning for a European plant commenced. It would be close to a carbon copy of the expanded Bayport plant. The Port of Rotterdam was offering major incentives for petrochemical producers to locate their plants in the city's new industrial park and so a Netherlands location was selected. It would essentially be a duplicate of the second Bayport plant, except built to European standards. It is interesting to speculate why Oxirane did not choose the PO/styrene option for the European plant. The fact that it would be simpler just to build a plant similar to Bayport may have been the controlling reason. Another may again have been the fact that Oxirane was not interesting in competing in the styrene

Fig. 11.3 Oxirane propylene oxide plant in Rotterdam. *Source* Lyondell Basell

market. The Dutch plant started up in 1972. Oxirane did not initially have a good outlet for tertiary butyl alcohol, but dehydrated some of it to isobutylene, which was sold or hydrogenated to recycled isobutene. Later, it made MTBE, a gasoline additive. Oxirane Netherlands was financed with European bonds, a windfall for Halcon and possible because of the tremendous success of the Bayport plant (Fig. 11.3).

The first PO/styrene plant was built in Spain for a firm called Montoro, a joint venture between Oxirane and Alcudia. It was sized at 32,000 metric tons per year of propylene oxide and 80,000 T/Y of styrene. (Somewhat later, after Alcudia had been acquired by Repsol, a Spanish petroleum refiner, the Montoro plant was substantially expanded, but without a license from Oxirane. A suit followed in the European Union court in Brussels, which was won by Oxirane. Repsol claimed that the experience gained by Oxirane for commercialization of the PO/styrene technology should be worth a license for its expansion, but this argument was rejected by the court).

Around that time, Halcon had become aware that Shell was building a PO/styrene plant with epoxidation technology at Moerdyck, also in the Netherlands. According to one of SD's lawyers, Barry Evans, Halcon had filed a number of patents on PO/styrene in 1969 and felt sure that Shell was infringing on its patents. Halcon sued Shell at the Dutch patent office, but was initially unsuccessful in obtaining relief. It appeared that Shell was relying on a patent using titanium catalyst rather than Halcon's molybdenum, although Halcon had also included titanium in its applications. After a frustrating several years of being turned down, Halcon prevailed when the Dutch patent office unexpectedly granted relief by giving priority to a simple, but very broad Halcon patent that covered co-production of PO and styrene. This action gave total victory to Halcon and Shell settled the dispute with a payment of $36 million, the price of Halcon's license to Shell for the Moerdyck plant. This also appears to explain why Shell never built a PO/styrene plant thereafter, in the United States.

Oxirane built another plant in Texas, this time at Channelview and with the PO/styrene technology.

Oxirane also entered a joint venture in Japan with Sumitomo Chemical and Showa Denko as a partner. The background is interesting, as the author found out in recent discussion with Ryoto Hamamoto, at the time a high level executive at Sumitomo Chemical Company and James Yoshida, who had formed and managed SDPlants, Tokyo for a number of years. There had been discussions between Mr. Yoshida and Mrs. Hamamoto and Mitsudo of Sumitomo Chemical about a license for SD's aniline process, but this turned out to be impossible as SD had already granted such a license to Mitsui Chemical. However, Yoshida said "how about Halcon/SD becoming 50–50 partners with Sumitomo Chemical in a large propylene oxide/styrene plant?" This immediately piqued Sumitomo's interest, as the firm had been studying the PO process. Both Mitsui Petrochemical and Mitsubishi Petrochemical were pursuing Halcon for a PO license at that time, but Sumitomo was apparently willing to enter an operating partnership, which Landau was extremely interested in.

According to the author's recent discussion with Hamamoto, Sumitomo Chemical was almost desperate having to find a propylene user, as at that time Japanese crackers, which were based on naphtha feedstock, were burning much of the byproduct propylene. Sumitomo had been unable to get a polypropylene license from Montecatini and so there was a real problem for the company to build a new naphtha cracker at Chiba, a very desirable location near the largest Japanese market. Sumitomo was convinced that it could not proceed without a large propylene user. So, the Halcon technology would be a Godsend for building the cracker, with two strong justifications. All of this made Sumitomo particularly interested in a deal with Halcon/Oxirane.

The plant was sized at 90,000 metric tons per year of propylene oxide and 225,000 T/Y of styrene. According to Yoshida, the size of the plant was set at 90% of Japan's installed chlorhydrin-based propylene oxide capacity on the assumption that all then existing Japanese PO producers would have to shut down, due to the new venture's breakthrough production economics. Amazingly, Oxirane gained its part ownership position by contributing only the PO/styrene technology as equity for its 50% ownership! Importantly, the first Oxirane PO/SM plant at Montoro, Spain was running very well by then, which is why Halcon could put such a high value on the technology. The plant started up in 1977, located adjacent to two other large new ethylene plants in Chiba. (A few years after that, when Arthur Mendolia (see later) was meeting with the Japanese partners in a joint venture board meetings in Hawaii, they spoiled the fun of an afternoon golf game when they complained that Oxirane was taking out large amounts of cash from the joint venture without ever having contributed anything but technology and startup help for its equity in the venture. An interesting sidelight on Landau's negotiating ability! When Hal Sorgenti became president of Arco Chemical, the Japanese were still wondering why Halcon had gotten such a great deal. Sorgenti said that he pacified them somewhat by shipping a few million pounds of free propylene oxide gratis to the Japanese partners).[7]

By late 1977, Oxirane was a relatively large company, in terms of assets and income and Halcon owned half of it. In his speech to the Newcomen Society in June 1978, Landau referred to a list of the 100 largest U.S. private companies in terms of revenues, with Cargill as No. 1 with $10 billion in sales and H. B. Zachry No. 100 with $200 million, claiming that Halcon fit in this group near the middle of the list. Landau went on to say that none of the other companies were in chemical manufacturing and marketing nor in high technology research.[8]

In the same speech, Landau talked at length about entrepreneurship and the ability of inventors to convert technical work into economic performance and their invention into a business. Quoting Peter Drucker, he referred to the ability to create and direct an organization for the *new*, the need for entrepreneurs who ask: where are the opportunities for a new industry, or at least for a new major process of product. In an age of rapid change, a technological strategy is essential for the success.......of a business...the market is the most potent source of ideas for innovation[9] (Fig. 11.4).

With Halcon now a more complex company, with a number of activities ranging from research and engineering to manufacturing and marketing, Landau and Rehnberg believed that they needed a seasoned chemical executive at its head. Landau had met a high level DuPont executive, Arthur Mendolia, in the early 1970s when Landau approached DuPont as a possible partner for a new ethylene glycol process then being researched by Halcon. However, no agreement had been reached on a possible deal, because Landau wanted DuPont to agree on an ethylene price, which Mendolia refused to do.

Mendolia then took a stint in the Defense Department, expecting to return to DuPont. However, Landau believed that Mendolia would be a perfect president of Halcon, with Landau remaining as chairman and chief executive officer. Mendolia at first resisted leaving DuPont, but when Harry Rehnberg suddenly died, the need became even greater and the offer presumably more attractive. Landau invited Mendolia to a lunch, his usual way to entertain and impress high level executives. Mendolia said that he could not refuse a lunch at Lutèce, a very tony establishment, and accepted the job. Mendolia also became CEO of Oxirane and moved to Princeton, where Oxirane's offices were located. This would be his main function at that point (Fig. 11.5).

Obviously, Landau's interest now was to expand the manufacturing part of Halcon—the main moneymaking part. When Mendolia talked to SD people, they worried that Ralph had abandoned them, because he was spending very little time with SD. The engineering subsidiary had six or seven hundred people, but it was fading into the background as excitement about Oxirane surged. The R&D work was being done in a separate subsidiary and so SD was basically an engineering company that had to find work for itself unless it received projects from Oxirane

---

[7] Harold A. Sorgenti. op.cit.
[8] Ralph Landau (1978). Halcon International. Op.cit.
[9] Ibid.

Halcon International, Inc.

An Entrepreneurial Chemical Company

DR. RALPH LANDAU

MEMBER OF THE NEWCOMEN SOCIETY
CHAIRMAN AND CHIEF EXECUTIVE OFFICER
HALCON INTERNATIONAL, INC.
NEW YORK CITY

THE NEWCOMEN SOCIETY IN NORTH AMERICA
NEW YORK   DOWNINGTOWN   PRINCETON   PORTLAND

1978

Fig. 11.4 Event by Newcomen Society honoring Ralph Landau. *Source* Newcomen Society in North America

or from licensing contracts—principally ethylene oxide—but these were becoming less frequent, particularly with strong competition from Shell/Lummus.

Mendolia joined Halcon when the new Oxirane plants were being built at Channelview. He received a clear awakening when he understood that as of that time, Oxirane had borrowed four hundred and sixty million dollars from a twenty bank

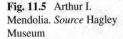

Fig. 11.5 Arthur I.
Mendolia. *Source* Hagley
Museum

syndicate headed by Chemical Bank.[10] Mendolia had never had to deal with such matters at DuPont, which were handled by DuPont's financial group. Now, Mendolias' main job was to keep the banks happy.

But it got more difficult. Ironically, this was again about the new ethylene glycol technology that Landau had approached him with several years ago, but now Landau had convinced Arco Chemical to invest in a new $130 million project for Oxirane that would build that MEG plant next to the Channelview PO plants.

Mendolia went about making Oxirane a real company in the traditional sense. He hired Don Wood, whom he knew as a competent Exxon executive, to set up a London office for Oxirane, which had operations in the Netherlands and in Spain and was selling PO in a number of other countries. With Halcon's research now in a separate company, Mendolia, in 1978, set up a laboratory for Oxirane at the new Princeton Forrestal center.

---

[10]Mendolia (1998).

Fig. 11.6 Halcon's new research center in Montvale, N.J. *Source* Newcomen Society in North America

Mendolia was also instrumental in starting a quarterly publications called *Columns* that was intended to keep employees and retirees up-to-date with news about Halcon. There was information about new research activities, interviews with key people in the organization, description of various activities, construction of new facilities, such as additions to the catalyst manufacturing plant, letters from foreign representatives and sales people, and announcements of new hires.

In October 1979. Mendolia hired Cy Baldwin, a long time high level executive of Stauffer Chemical Company to become president and chief executive of Oxirane. In a long interview in *Columns* published in the fall of 1980, he discussed the work that Oxirane's research group was doing on an SD process to make methyl ethyl ketone (MEK) from tertiary butyl alcohol, as produced at Bayport. He also announced a joint venture with FMC Corporation for the manufacture of glycerin at Bayport, using SD technology.

In the interview, Baldwin was asked about the status of the new ethylene glycol plant that was currently shut down because of operating problems. Baldwin said that research was being done to try to solve these problems.

This was the status of Halcon near the end of 1980. Scientific Design had reorganized itself to become an operating company with research and engineering subsidiaries. Through Oxirane, it received a very substantial income from its highly successful propylene oxide process. Its Scientific Design subsidiary was a relatively large engineering firm with a large overhead. And Halcon's research arm was continuing to develop new technology and had built a large new research center for 200 people in Montvale, N.J. (Fig. 11.6).

In 1979, Landau reached out to Robert Malpas, a recently retired ICI main board member, and invited him to become Chief Executive Officer of Halcon, with Landau remaining as Chairman. Mendolia became chairman of Oxirane with Cy Baldwin as president. Everything seemed primed for a great future. However, Halcon had incurred massive debt and the new MEG process started to look like a failure. The story continues in Chap. 12.

References

Arthur I. Mendolia (1998) Oral History. Chemical Heritage Foundation. Philadelphia, Pa.
Harold A. Sorgenti (2003) Oral History. Chemical Heritage Foundation, Philadelphia, Pa.
Haldor F.A. Topsoe (1999). Oral History. Chemical Heritage Foundation. Philadelphia. Pa.
Ralph Landau. (1978). *Halcon International, Inc.* Speech to Newcomen Society of North America.
Ralph Landau. (1994) *Uncaging Animal Spirits*. The MIT Press. Cambridge, Mass.

Chapter 12
A Reach Too Far

Abstract SD's laboratory had, for several years, been developing a new technology to make ethylene glycol directly. Based on the success of Oxirane, Landau convinces Arco Chemical to join in building a large new plant based on the new process. However, no large scale pilot plant work had been done to prove out the rather complex technology nor the construction materials needed for the highly corrosive parts of the process. After a year of fruitless attempts to solve various problems, the plant is scrapped. Halcon, saddled with very high debt for financing its share of Oxirane, is forced to abandon its partnership in the Oxirane venture and sells the technology and its Oxirane partnership to Arco, receiving several hundred million dollars in the transaction. After some money was paid to other Halcon shareholders, Landau and Rehnberg's estate receive a payout allegedly amounting to several hundred million dollars.

The new monoethylene glycol (MEG) plant that Halcon and Arco Chemical built as a joint venture project at Channelview Texas turned into a disaster! The plant did not work well, suffered extreme corrosion problems, and had to be scrapped. This in itself was bad enough, but the consequences were even worse. Because the plant could not deliver the large amount of contractually promised MEG, it was necessary to go into the market to buy MEG at high prices adding up to $50 MM to meet commitments to clients. And because Halcon, which was already in substantial debt to finance its share of the new Channelview propylene oxide plant, was in even more debt to the banks for its share of the MEG financing. Halcon's financial position was so dire that the only way out was to sell its share of Oxirane to Arco Chemical and get out of the global propylene oxide business. And so, Landau's dream to become the CEO of a large manufacturing firm was never realized and Landau went into retirement not many years after matters were settled.

It is ironic that the chemical that can be said to have "made" Scientific Design Company, ethylene oxide, the traditional source of MEG, also became the reason for Halcon's downfall. It is a story with almost Shakespearean overtones, a case where the previous model for the company's success was overtaken by *hubris*. How and why the MEG project failed is the subject of this chapter, which also covers the buyout.

© Springer Nature Switzerland AG 2019
P. H. Spitz, *Primed for Success: The Story of Scientific Design Company*,
https://doi.org/10.1007/978-3-030-12314-7_12

12.1   Background of the MEG Project

Over the time between the startup of the first SD EO plant, Naphtachimie and the early 1970s, SD had licensed over a hundred EO plants worldwide in serious competition with Shell/Lummus. Since Shell had developed an EO process based on oxygen rather than air, SD countered by developing its own version of this technology, able to get around Shell's patents. Oxygen-based plants had a lower capital cost than air-based plants, particularly if the oxygen was brought in by pipeline rather than made as part of the EO plant. The oxygen processes raised the EO yield from 71% to around 73%, with the Shell catalyst considered slightly superior to the SD catalyst. SD was able to offer either the air- or the oxygen option and could convert its air-based plant into oxygen-based, thereby obtaining more capacity. Meanwhile, Union Carbon and Dow continued to expand their plants, with Dow continuing to use the chlorohydrin technology in the U.S., but licensing SD technology at Stade in Germany and in Spain. Royalty income gained by Halcon/SD was decreasing, as the other technologies (maleic anhydride, cyclohexane oxidation, isoprene) were not contributing much and the profit on EO plant design and licensing was increasingly diminished as a result of competition with Shell. Other companies, principally in Asia, were developing their own version of an ethylene oxide process.

So, with licensing income declining, no new technology ready to be commercialized, but with the glorious success of Oxirane, Landau's thinking was quite clear. He had always wanted to create and head up a manufacturing enterprise based on original chemical research. He recalled that in the early 1950s when SD's work on ethylene oxide technology appeared to be commercially viable, he had approached Sears Roebuck, which was selling large amounts of antifreeze, and Shell Chemical, which had surplus ethylene, to form a joint venture to manufacture ethylene glycol. This bold plan never materialized and Shell decided to start its own research on ethylene oxide.

But now he was very close to achieving his dream. Landau had always known that the value of a company that received income from engineering and licensing would never have a market value close to approaching that of a company that engaged in manufacturing. Selling the company was always the ultimate goal, but first it was necessary to build substantial value into the firm and that could only be done by increasing the amount of income received from manufacturing chemicals rather than from licensing its technology. Now he had succeeded by having his firm, Halcon International, become a 50% owner of a highly successful propylene oxide joint venture with Arco. The next step was clear. Halcon needed another process that would lead to part- or whole ownership and, together with the income from Oxirane, would make Halcon a large chemical company with income primarily from manufacturing.

In terms of domestic uses of ethylene oxide, traditionally almost two-thirds of the chemical goes into the production of polyester fibers and anti-freeze. Non-ionic surfactants, which are produced by reacting ethylene oxide with a long chain natural or synthetic alcohol, are the next large consumers (Table 12.1).

Table 12.1   U.S. ethylene oxide markets in 1980

	MTons/year	% of total
Monoethyhlene glycol	1,645,000	63.9
Nonionic surfactants	278,000	10.8
Heavier ethylene glycols	276,000	10.7
Glycol ethers	162,000	6.3
Ethanolamines	162,000	6.3
Other	52,000	2.0
Total	2,375,000	100.0

Source SRI Chemical Economics Handbook

A process to make ethylene glycol directly at high yield would be a winner. Such a process would theoretically have a yield 20% higher than the route to make ethylene glycol from ethylene oxide based on the traditional silver catalyst. SD had been highly successful at developing new oxidation technology "from scratch" and so developing such a process was considered a reasonable target.

In an attempt to develop a leadership position in ethylene glycol (MEG) manufacture, Landau and Dave Brown had some time ago instructed Halcon's research group to look at the possibility of a catalytic process that would produce ethylene glycol directly. Specifically, John Kollar, who had spearheaded the research and development of the cyclohexane and epoxidation technologies, was charged with finding a direct route to ethylene glycol. In this he was joined by, among others, John Schmidt and Bob Hoch, both highly experienced researchers and engineers. Serious work commenced around 1971. Kollar had actually filed in March, 1969.

The approach that had looked the most promising to Kollar was a technology known as acetoxylation. In such a process, ethylene is reacted with acetic acid to form glycol mono- and di-acetate (1,2 diacetoxyethane). This is then hydrolyzed to ethylene glycol and the recovered acetic acid is recycled. A number of other companies were working in this area, including Celanese, DuPont, Teijin and Kuraray.[1] Kollar believed that the most relevant work was out of ICI on a vinyl acetate process, which however, had failed because of serious corrosion problems. Kollar modified the oxidation conditions to synthezise glycol acetates instead of VAM. Selectivities of up to 98% were reported for the first step. The catalyst Kollar eventually decided on contained tellurium and bromide and the reaction was carried out around 400 psig. Given the acetic acid and bromide present in different parts of the process, it was evident that there would be highly corrosive conditions, requiring specialized construction materials. Another issue involved the fact that the hydrolysis of the diacetate involves high utility consumption, as the glycol is recovered in dilute solution.

Readers interested in the details of Halcon's process are referred to the following patents, among others:

---

[1] Weissermell and Arpe (1978).

U.S. Patent 3,715,389   Robert Hoch, John Kollar
U.S. Patent 3,872,164   John P. Schmidt
U.S. Patent 3,907,874   Robert Harvey, John Kollar, John Schmidt
U.S. Patent 4,245,11   Alan Peltzman

In 1972, Landau had approached Arthur Mendolia, then a high level executive at DuPont and told him that Halcon was developing a direct ethylene glycol process and was interested in a partnership with DuPont for the commercialization of the new technology. DuPont was logical, as the company was buying large amounts of MEG for its *Dacron* polyester business. Landau wanted DuPont to supply the ethylene needed for the process on a long term basis at an attractive price. This possible venture fell apart when the first oil crisis in 1973 made future ethylene pricing uncertain.[2]

By 1975, the laboratory had carried out considerably more work on this new MEG route. When the yield started to look attractive, a small pilot plant was built and engineering studies were commenced. A major hurdle was the fact that the first plant would have to be quite large to achieve attractive scale economics, as the conventional process plants would otherwise have scale advantages over a small new plant with a new process. It was decided that the plant needed to have a capacity close to a billion pounds per year—actually the design was for an 800 million pound U.S. plant, which was originally estimated to cost approximately 130 million dollars. It had a rather unusual flowsheet and the materials of construction were expensive due to the chemicals involved.

What kind of economic advantage would a new process based on the acetoxylation technology command? This question would be examined over the next period as the estimated capital cost of the plant became known, the utilities consumption calculated, and the yield confirmed. It would not be a "world beater", but it would be better than the traditional route.

Arco Chemical, Halcon's partner in Oxirane, was following this closely and became interested in partnering again, making this another part of Oxirane. It was said that Bob Bent, Arco Chemical's president, strongly pushed Landau on this, but was first rebuffed. Landau allegedly did look for other partners, including ICI. When this did not materialize and the estimated cost of the plant soared to 178 million dollars, Landau agreed to make the project another joint venture with Arco. The higher-than-estimated cost of the plant would, however, further narrow the economic advantage that the new technology would have over the conventional process. Some studies were carried out to try assess the economic advantage of the new process. Halcon President Malpas asked Brian Ozero, who was working for him at that time, to develop a report on the ultimate potential of the conventional EO process. According to Ozero, the report showed that improvements in the conventional process could reach economics that would come close to the new technology. However, Malpas did not let this report stop the venture, most likely because of Landau's adamant push for the new EG process (Fig. 12.1).

---

[2]Mendolia (1998).

Fig. 12.1 Recent Picture of
Sir Robert Malpas

Speaking to SD engineers who helped on this book, it became clear that somewhat of a rift had developed among the various participants to this decision. Bob Hoch, who was part of the research group, remembered that the monthly reports prepared by John Kollar, for a long time stated that the technology was not yet ready for commercialization, at least in the form that the plant was currently being designed. It is not known whether Kollar still felt that way when it was decided to go ahead and build the plant. Kollar's reservation may have stemmed from the fact that the pilot plant that was being operated was quite small in terms of stream volumes, which made it difficult to judge the amount of byproducts, particularly heavy "bottoms" that formed in the first step. The heavies were originally believed to be all higher glycol acetates, which would equilibrate on recycle. Kollar's concern was probably based on his recent findings that part of the heavies were not recyclable and would have to be removed. The commercial plant had no provision for such removal. There were the Halcon people who had not participated in the success of Oxirane and saw the MEG technology as their chance for a big win. These were mostly engineering people who were not particularly familiar with the process. These people were pleased that Landau was pushing strongly for the project and could not be dissuaded. On the other hand, several experienced engineering executives were very skeptical. Harold Huckins remembered walking into Landau's office and giving him ten reasons why the plant should not be built. at least as currently designed. He was summarily dismissed for his negative views. Harry Peters, Vice President of Engineering had recommended to Landau that a thirty percent contingency should be built into the estimate. This also fell on deaf ears. Then, there was the question of which SD engineers should design this first-of-a kind plant that would require great chemical engineering skill. Neil Yeoman who, with Brian Ozero, had done much of the design of the Oxirane plants, was working on the design of the new Channelview plants and was not available. However, Ozero, considered by many SD's top design engineer, was available and was interested. However, when he was told he would be on a team headed up by Harold Gilman, Ozero declined. Gilman, who was in Hoch's opinion as good a process engineer as you could get and so regarded by Landau, remained the head of the team, reporting to Dick Kaplan, a very competent group leader.

There is reason to believe that Arco Chemical did not spend a lot of time looking at Halcon's work on the MEG technology. Arco Chemical's research people were heavily engaged in developing a new environmentally favorable isocyanate process for urethanes manufacture and left much of the engineering thinking about the MEG plant to Halcon and to people at Oxirane. A recent discussion with Wayne Wenzheimer, then a research director at Arco Chemical, revealed that on new technologies, such as the isocyanate process they were working on, they were using a large pilot plant with a large dedicated staff to develop enough information to make scale-up to a commercial plant less of a risk. Spending little time with Halcon on the small MEG pilot plant, they were skeptical about the idea of designing an 800 million pounds per year first-of-a kind-plant with information only from a very small pilot plant. However, Arco recognized the success that SD had achieved with similar very large scale-ups and evidently put their trust in SD/Halcon. Interestingly, Arco's extensive pilot plant work resulted in the decision to cancel the isocyanate work. It was found that it was impossible eliminate totally the impurities in the final product which were attributed to the selenium catalyst employed.

The financing of the MEG plant created an additional big financial burden on Halcon, which was possibly still in debt on the expanded Bayport plant. It had to borrow money for the Channelview PO/SM plant and now needed to borrow more. An attempt to develop a picture of Halcon's obligations relative to its position in Oxirane looked as follows:

- The debt of Halcon to Arco Chemical for the Bayport plant and expansions involved non-recourse loans that were being repaid with a series of notes. It is impossible to know whether this loan was completely repaid by the time the MEG plant went into operation.
- The plant in Netherlands was financed by Oxirane, using offshore bank financing so no problem.
- Halcon did not put any money into the Japanese Oxirane joint venture with Sumitomo. The equity was in the form of a royalty-free license. There was no debt obligation for Halcon. There was similarly no financial obligations with respect to the Montoro plant in Spain.
- The Channelview PO/SM plant was financed with Arco contributing its 50% in the form of guaranteed bank debt and Halcon using bank debt also for its share, with a floating interest rate relative to prime. At the start of borrowing, the prime rate was 7%, but was starting to rise rapidly as a consequence of the second Oil Shock.

The MEG plant was project financed, meaning that the banks made payments to contractors when successive progress targets were met. The loan was guaranteed by Oxirane, which had a high cash flow. Contracts for prospective product sales were also considered by the banks as supporting the financing. The loan for Halcon's share of the MEG plant was also at a prime rate of around 7%. Bent was risking a lot of Arco's money, but the parent company was starting to get a lot of income from its Alaskan North Slope position with oil flowing down the Alaskan pipeline and to California and other refineries.

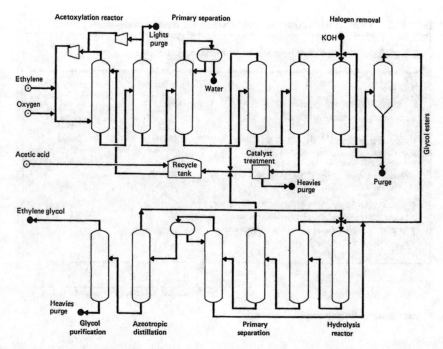

Fig. 12.2 Union Carbide conception of Halcon's MEG process from patents. *Source* Chemical Engineering

According to Arthur Mendolia, then president of Oxirane, the company, in 1980, owed 460 million dollars to various banks.[3]

Considering what happened soon after the startup, it is interesting to look at an article published in 1978 that quoted Art Brownstein, a well-known petrochemical expert who was then New Ventures manager for Exxon Chemical. "On paper, the Halcon process is simple," he said. The initial step of olefin bromination is followed by acetic acid displacement of the bromine to yield ethylene diacetate, plus monoacetate due to some partial hydrolysis. Displaced bromine is reoxidized by tellurium, which itself is reoxidized by oxygen. The trouble with this route is that it is costly to recycle acetic acid. You must distill to get the water out, and this takes a lot of energy. And running the reaction in glacial acetic acid creates a corrosion problem, so that investment cost is high.[4]" Brownstein had said at an AIChE meeting in Houston, Texas in 1975 that an acetoxylation route to ethylene glycol was being researched, that it would be attractive if commercialized, but that eventually, the most economic route would be one that was based on synthesis gas. (It was actually commercialized in China some forty years later, based on gasified coal)[5] (Fig. 12.2).

---

[3]Ibid.

[4]Chemical Engineering. Jan. 15, 1978. 67–69.

[5]Chemical & Engineering News (1975).

Table 12.2   Union Carbide's comparison of Conventional versus Acetoxylation process

Ethylene use and energy balance of competing processes		
	Conventional	Acetoxylation
Ethylene use, lb/lb of glycols	0.646	0.486
Energy consumed by process,* Btu/lb of glycols	7,280	11,600
Byproduct energy,[†] Btu/lb of glycol	5,660	2,200
Net energy supplied by utility systems, Btu/lb of glycols	1,620	9,400
Gross energy used by utility systems[‡] Btu/lb of glycols	2,310	13,400

*Process heat losses, steam for separations, etc.
† Available from reaction heats and residue-fuel values
‡ At 70% overall efficiency in the utilities area
Source: Union Carbide Corp.

Source Chemical Engineering

The Brownstein article also included comments on the new Halcon process by Union Carbide ("High energy consumer") and by Celanese ("We scrapped a plant [actually a new process for vinyl acetate…Ed.] a few years ago because of serious corrosion problems"). Union Carbide had evaluated the Halcon technology, comparing it to the conventional process still in use by Union Carbide. The comparison showed substantial yield advantages, but massive energy use for the Acetoxylation process (Table 12.2; Fig. 12.3).

Did Landau see this article? This could have been a premonition!

12.2   Plant Startup and Operation: Serious Problems

The MEG plant was designed based on pilot plant operation and extensive engineering studies and was placed in operation in the middle of 1977. The startup manager was Roger Cox from SD, assisted by Arco senior manager Eric Andersen. Jon Valbert, who had run the MEG pilot plant, was also on the startup team. The plant started up quite well, with no serious problems. Harold Huckins recalled that he personally sampled the ethylene glycol product when everything was operating well. He said that the sample looked completely clear and met all the specs. This showed that the design of much of the plant was, in fact, sound. Alan Peltzman, who worked under Kollar, also confirmed recently, that he had noted specification glycol coming out of the back of the plant.

Fig. 12.3 Photograph of the MEG pant under construction. *Source* Harold Huckins

However, after only a couple of weeks, a stoppage occurred in piping around a recovery column. At the bottom of this column there was a circulation system, which was like a natural draft reboiler which withdrew a stream and reinjected it into the column while withdrawing a bottoms purge stream. The circulation system did not work well, leading to plugging. Tellurium catalyst was present in the stream which also contributed to the plugging. Moreover, a lot of the expensive tellurium was lost in the purge stream, though the exact way the system operated or was supposed to operate is unclear.

Then, another, serious problem surfaced. During the frequent shutdowns, it was possible to inspect different sections of the plant. Some corrosion problems had been anticipated so that certain parts of the plant were designed with HastelloyB and titanium. So, it was a great shock that within literally a month, serious corrosion occurred in different areas, including some columns. In one case, it was decided to install Teflon sieve trays instead of the titanium trays initially provided. Since tantalum has even better corrosion resistance to acetic acid and bromine, some of this even more expensive material was used in replacement.

There was a period of about nine months when the startup crew and others were trying to solve these problems.

Around this time, John Schmidt, who was then Executive Vice President of Oxirane International, was called to the MEG startup by Landau and became the "czar" for solving the problems. He and Huckins, who was Vice President of Technical Operations, spent full time trying to help solve the bottoms plugging problem, which seemed to have a lot to do with the poor circulation from the "reboiler". The fundamental process problem was the production of substantial amounts of heavies and solids not anticipated in the design causing blockage in lines, pumps, and heat exchangers, as well as unacceptable purges of organics and tellurium. Valbert was asked whether there had not been undesirable solids formation in the pilot plant. He said that they had indeed encountered this problem and had added more bromine to overcome it. The commercial plant could not have been operated that way.

A number of things were tried. If less tellurium was used, the reaction rate decreased, but there were fewer heavies and so the bottoms problem improved somewhat, but the yield dropped. When Landau insisted that more tellurium should be injected to meet the design yield conditions, this made the plant inoperable. Other variables, such as oxygen levels, pressure and temperature were varied. No solution was found. Some equipment changes were also made over this period.

Corrosion strips had been installed initially in various areas and more were installed as time went by. The plant had been operated for over a year when, during a shutdown, it was discovered that the titanium corrosion strips installed in the metal wall of a large column containing acetic acid and HBr showed serious thinning, as well as embrittlement attributed to hydrogen diffusion. There was concern that this was an extremely serious problem, as a "catastrophic" failure of a vessel could occur, with potentially fatal consequences for people who might be in the surrounding area (Fig. 12.4).

It is hard to know whether the corrosion discovery or the inability to solve the plugging problem led to consideration to not to start the plant up again after the next plugging incident. By September 1979, after two years of startup attempts, the operation was losing $4–5 million per month. It was felt that the long term outlook was not good even under the best assumptions of (a) willingness to spend a lot more money, (b) success at solving many operating problems rapidly and (c) ability to achieve much higher yields than the plant had demonstrated. No information is available on estimates on what it would have cost to use tantalum in a number of places where titanium was attacked by the corrosive mixtures in different parts of the plant.

Mendolia asked his engineering group to evaluate the MEG plant. It was their conclusion that the process would actually not be more economical than the conventional process (for ethylene oxide, converted to glycol) because the MEG process was a big energy user and energy costs had gone up a lot faster than ethylene costs. (!) This finding was undoubtedly more significant to Arco Chemical than to Halcon, since Arco would think that they were backing a losing process, while Halcon had all its eggs in the MEG basket.

In November, Mendolia announced that the plant was being shut down.[6] There was obviously tremendous consternation, particularly among the Halcon people, since failure of the plant would have a similarly catastrophic effect on Halcon, with

---

[6] Arthur I. Mendolia op.cit.

Fig. 12.4 Harold Huckins posing at MEG plant. *Source* Harold Huckins

its huge debt. For this reason, Landau was adamant that Halcon would not agree to giving up on the startup, leading to a permanent shutdown. He said that there was no reason to think that both problems, bottoms plugging and corrosion, could not be solved. So now the MEG partners were at odds. For Halcon, everything depended on getting the plant to work. For Arco Chemical, which had financed Halcon all the way, based largely on Bob Bent's decisions, there was little reason to spend more money for an indeterminate goal. Because of Halcon's high debt, Arco knew it could buy Halcon out and become the sole owners of Oxirane and the only question was the price. A big negotiation was ahead.

A meeting was called by Arco Chemical to decide on scrapping the plant. Landau allegedly refused to attend.

Why did things go so badly for this plant and the process? In an earlier chapter the SD philosophy was clearly set out. For all of its successes, SD had usually "oversold" a process at an early stage, assuming (in all cases correctly) that the problems that would inevitably come up could be solved with time, whether in the pilot plant or in the actual first of a kind plant. Egbert made the first ethylene oxide plant work. SD was able to make the mixed acids plant at Joliet work by converting it to make only terephthalic acid, "pulling a rabbit out of the hat" for Amoco, which had opted for mixed acids in the plant design. The chlorinated solvent plant for Pechiney at St. Auban did not work and important materials of construction substitutions were required to deal with serious corrosion problems. SD was unable to make the Monsanto cyclohexane oxidation plant work, was sued by Monsanto, but settled and then licensed seven large nylon intermediates plants. The isoprene plant did not work until the cracking step catalyst was changed from bromine to ammonium sulfide. On all these cases, insufficient pilot plant work nor semi-works demonstration units were employed, requiring SD to make changes to the plants when initial operations demonstrated that not enough information had been collected to come up with an operable first-of-a-kind plant. Only the epoxidation technology worked well in the first plant, built at Bayport, most likely because Arco Chemical had run a large pilot plant on the PO/TBA version, the one used in the Bayport plant. Still, the first PO/SM plant, built in Spain, also did work well and was based on Halcon's small scale pilot plant work.

Over the thirty years of SD's success, Landau's approach had been to persuade company partners to put up money based on early SD laboratory success, figuring that SD's researchers and engineers would achieve the promised yields in a working plant even if changes had to be made along the way. There was no other way for SD to operate, considering the circumstances. The company simply did not have the resources that allow large companies to develop technology stepwise and more slowly in a more classical, measured and costly way.

This time, the approach did not work!

Finally, the question on the minds of many of the people involved was: how good could this technology be? This was actually never resolved to everyone's satisfaction. If the process could be run successfully at the highest laboratory yields obtained, if the utilities consumption could be lowered, for example by selling byproduct steam and if the price of energy came down to what it was when the plant was first conceived, if the very expensive tellurium catalyst in the waste streams were recovered and if the needed, very expensive construction materials, such as tantalum, were used to replace some titanium without occurring prohibitive capital costs, then a new plant with all these modifications might still have a small advantage over the conventional process. Most people would say "too many ifs!" so that the general conclusion was that even a more classical development program with a much larger pilot or semi-works plant most likely would not have produced a "world-beater" technology that would have replaced the traditional ethylene-oxide-based silver catalyst technology that has remained the process new producers are still choosing at the time of this writing.

Historical Prime Rate Graph

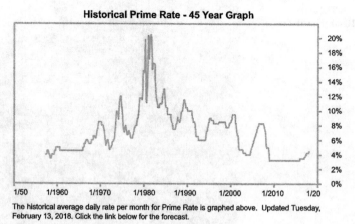

Fig. 12.5   *Source* U.S. Commerce Department publication

12.3   The Oxirane Buyout

When Arco Chemical's management decided not to put any more money into the MEG plant, Halcon's ownership of 50% of Oxirane was doomed, as the financial burden on Halcon would have forced the company into bankruptcy. It is difficult to reconstruct the financial facts at that point, but the following seems quite clear:

– There was a huge bank debt. Halcon's interest payments for its share of the Chan-nelview PO/SM plant had risen from around 7% to the 15–20% level (see graphic below) as a result of the second oil shock and its aftermath. This would require interest payments more than twice as high as anticipated. And then there was Hal-con's debt for its share of the MEG plant, probably in the range of $125 million, which would have to be repaid (Fig. 12.5).
– Also, Oxirane had incurred the payments to satisfy contractual commitments for ethylene glycol it could not deliver to customers at a time when glycol was very tight. The reason for tightness was that other glycol producers had not expanded production when Oxirane announced the construction of an 800 million pound plant.
– Hal Sorgenti recognized that Arco Chemical itself was having problems meeting the debt commitments and interest payments for its share of Oxirane.[7]

There is a poignant anecdote! Bob Anderson, the legendary chairman of Atlantic Richfield and a "dyed-in-the wool" oil field wildcatter, visited the stricken plant. He then called Hal and said "It's a dry hole. Plug it".[8] Spend no more money and get

---

[7]Sorgenti (2003).
[8]Ibid.

Fig. 12.6 Harold I. Sorgenti
(Artist sketch). *Source*
Science History Institute

out. So, Sorgenti, who had only recently attained his position, studied all the Oxirane contracts and finances and then went to see Bob Malpas, then chairman of Halcon and someone Sorgenti knew quite well. He asked Ed Muller, then Halcon's chief financial officer, "Could you explain to me where you put money into this (Oxirane) venture because I don't see any place where money comes in from you, it only goes out to you. You get a share in the profits and the dividends, but you never put any money in" and the response was "Of course. We don't have any responsibility to put money in. That's why we did the joint venture with you". (laughter) And Sorgenti then said, "I'm here to tell you that yesterday our board said they weren't putting any more money into Oxirane. The plant is draining money from us, so we want to shut it down, write it off, and tear it down" (Fig. 12.6).

DuPont came to look at the plant at the instigation of Landau, who had talked Ed Jefferson, CEO of DuPont, telling him that "Arco Chemical just aren't good operators. They don't know how to run anything and they say it's a worthless plant. I told them we'll bring in somebody to take it over and run it." Jefferson called Sorgenti, who told him that the technology was faulty, that it was based on a very small pilot plant, and that Arco had already put in an extra 100 million dollars to try and make the plant run. "You're free to look at it. If your conclusion is different from ours, that would be great".[9] A DuPont team spent a couple of weeks at the plant, told by the firm's management to determine whether DuPont should consider buying the plant and making it run. This would conceptually have made sense for DuPont, since it was a very large buyer of ethylene glycol for its Dacron business. If DuPont had bought the plant and the Halcon technology behind it, Halcon might have been saved and might not have had to sell all or part of its share of Oxirane. An interesting thought! In any case, the DuPont people came to the same conclusion as Arco Chemical. This was the final blow to Landau and Halcon.

Sorgenti's final comment on the shutdown was that "the company had the world's largest supply of scrap titanium".

Shortly thereafter, Sorgenti went to a board meeting of Atlantic Richfield in California. At one point, Bob Anderson asked, "what shall we do about Oxirane?" Bill Kieschnick, then No. 2 at the parent company said, "the problem's being handled. Ralph refuses to put any money into it and Arco's not going to do anything either. The banks will call the loans. Halcon will shortly have to call us to bail them out". Anderson asked Sorgenti what will happen then. He said, "we're going to destroy a great company. Ralph will sue". Hal then said to Bob, "Call Ralph, meet him in New York for breakfast tomorrow and tell him we will buy him out of Oxirane. I know he has been talking to Felix Rohatin of Lazard Freres (& Company) and he's trying to sell his shares to him. I really think this is the time to go in and buy out Ralph's share, take (Oxirane) over and develop this wonderful franchise that we have (on propylene oxide)".

And that's what happened. Amazingly, Anderson and Landau quickly agreed on a deal for Oxirane, and the parties, including Lazard, started to work on it. First, there were severe tax implications for Halcon/Landau with 100 million dollars' worth of taxes the way the deal was originally negotiated. Anderson balked at changing the deal and so Rohatin said they would go to the Saudis if Arco would not compromise on the tax issue. Nobody believed that. Eventually, Anderson relented. Then Landau wanted another ten million dollars on top of the 250 million dollars Halcon was going to get to turn over Oxirane and the entire propylene oxide technology to Arco. Andersen became furious and was ready to balk again when Sorgenti said we will ask Ralph for other Halcon technology nominally worth ten million dollars. This was quickly agreed by Anderson, a package of patents was put together by Halcon and the deal was done.[10]

When all the papers were signed, Halcon had been saved, but it was no longer in the business of manufacturing chemicals.

The company continued for several years, but it was never the same again. The last chapter covers the five or so years when Halcon/SD continued to operate in the mode that had made Scientific Design successful as an engineering and construction form, though without the benefit of successful research leaders. Two years later, the company was sold to Texas Eastern Corporation for a reported 45 million dollars.

---

[9]Ibid.
[10]Ibid.

In 1983 it was time to "cash out. There was the sale of Halcon's half of Oxirane for the 260 million dollars from Arco. Then, there was money from the sale to Texas Eastern. In addition, there would have been some accumulated royalties from other technologies that had not yet been paid out along the way. Information made available to the author from one source provided the following: (a) when Harry Rehnberg died, the stock distribution became 2/3 Landau and 1/3 Rehnberg family estate. (b) Ralph Landau allegedly received about $180–200 million, the Rehnberg family about $90–100 million and other shareholders divided about $70 million. This would add up to around $350 million. These numbers could not be verified from other sources.

References

Arthur I. Mendolia (1998). Oral History. Chemical Heritage Foundation. Philadelphia, Pa.
Chemical & Engineering News. *New Processes eyes for Ethylene Glycol* March 31, 1975.
Harold A. Sorgenti (2003) Oral History. Chemical Heritage Foundation. Philadelphia, Pa.
Weissermell, K and H.J. Arpe (1978) *Industrial Organic Chemistry*. Verlag Chemie Weinheim-New York.

Chapter 13
An Ending

Abstract Halcon soldiers on. A new acetic anhydride process is sold to Eastman Chemical. The very promising methyl methacrylate process was never commercialized, as no financial partner could be found. SD continues to work on engineering projects and also starts an activity in environmental technology. However, there is not enough income from engineering work to sustain the firm. Since SD was still selling ethylene oxide plant licenses, designs and catalyst, the company is sold in 1987 to Texas Eastern Corporation and Landau retires. Scientific Design Company continues to exist as a source of ethylene oxide technology, competing as before with Shell/Lummus and foreign firms.

The chronology covering the dissolution and sale of Halcon is covered in a short memo made available by Neil Yeoman, who stayed with the company until the end.

1979: The ethylene glycol plant is shut down. Successful startup was never accomplished.

1980: Halcon sells its half of Oxirane to ARCO.

1981: The company was restructured for tax avoidance problems, a need that developed from the large cash inflow that accompanied the sale. It was renamed "The Halcon SD Group".

1982: The company was sold to Texas Eastern Corporation. Landau retired.

1985: Texas Eastern dismantled two of four divisions, the Engineering and Construction capability (Scientific Design) and the computer capability (Halcon Computer technologies). R&D, Process Licensing and Catalyst Manufacturing were retained. Company size was reduced from 600 to 200 people. The New York office was closed.

1986: Halcon Research and Development was dismantled. Company size was reduced to about 70 people, all involved in licensing and catalyst manufacture.Ethylene Oxide/Glycon/Glycol ethers and Maleic Anhydride technology were essentially all that remained.

1987: The company was sold to Denka Chemical, including the name "Scientific Design", which again became the name of the firm due to name recognition.

(The company was sold again three times, first to Mobay, later to Linde, and finally to a joint venture between Saudi Basic Industries and Clariant).

© Springer Nature Switzerland AG 2019
P. H. Spitz, *Primed for Success: The Story of Scientific Design Company*,
https://doi.org/10.1007/978-3-030-12314-7_13

This chronology paints a stark picture of what can happen to a company when a disastrous, financially untenable event results in the effective end of a highly successful enterprise. While its leader, Ralph Landau, perhaps understandably, lost interest in the firm as a result of the MEG fiasco, the talented group he left behind tried to keep the dream going.

With Landau, Saffer, Dave Brown, Gillespie, John Schmidt, and John Kollar gone and Halcon/SD now a subsidiary of Texas Eastern Corporation, it was a different company now, but still looking at the chemical industry for new work. Among the people remaining were Jon Rehnberg, chairman, Ed Muller, president of Halcon, (formerly top financial person), Bill Long (Patents and Licensing, Ted Stein (R&D), Joe Porcelli (Technology) and Ernie Korchak (President of SD).

The company boasted of its many successes in providing the technologies used in the production of a number of chemicals, including multi-billion pound 1980 global production capacities for ethylene oxide, propylene oxide, and terephthalic acid.

Unfortunately, the years under Texas Eastman were not particularly productive. Lacking the guiding skills and salesmanship of the founders and early hires, the company did not announce any new technologies (except, see below) and found that the existing processes, notably ethylene oxide and derivatives, did not keep the large engineering staff very busy. SD had always had to generate work from companies who had proprietary technologies which SD could help in commercializing or from companies who needed conventional engineering/construction services. This had become much more difficult. On top of that, the early 1980s were a particularly difficult period for chemical companies, which were suffering from the effects of the "Reagan Recession". They turned inward and stopped new projects. This turned out to be the final straw for a company that had thrived on sponsored innovation, but found no takers at a time when chemical firms experienced little or no profitability. And so the brave attempt to keep going foundered, partly as a result of external factors. When Texas Eastern itself got into difficulties, it was the end of Halcon.

13.1 Acetic Anhydride: The Last Success

At some point, probably in the late 1960s or early 1970s, John Kollar became interested in carbonylation chemistry, which had led to great success for Monsanto with acetic acid. Kollar was still thinking about nylon intermediates chemistry and thought that it should be possible to make adipic acid directly using this type of chemistry, which is generally based of using cobalt iodide or a noble metal as a catalyst, carbon monoxide and high pressures. At one point during a series of experiments, he asked whether there might be any interest in producing acetic anhydride using such an approach. He was carbonylating methyl acetate with a catalyst comprising rhodium chloride and chromium hexacarbonyl and an acetic acid solvent. Methanol is first esterified to the methyl acetate.

Halcon's patents came to the attention of Eastman Chemical, which was a major producer of acetic anhydride for the production of cellulose acetate. The technology

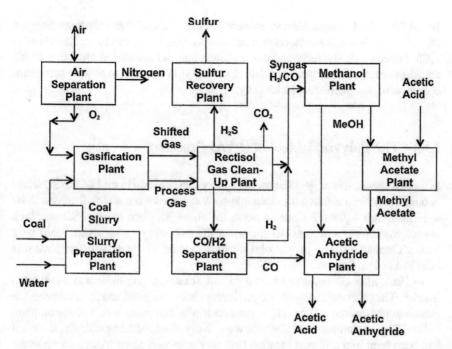

Fig. 13.1 Eastman Chemical/Halcon acetic anhydride process

it had employed for a long time was the Wacker process that cracked acetic acid to the anhydride. Eastman had been basing their operations on methanol produced in the conventional manner from natural gas. But when Eastman's natural gas contract was cut off due to a natural gas shortage in Kentucky/Tennessee, the company decided to make methanol from coal, which not only was abundant in the region, but was already being used to stoke Eastman's boilers for the production of steam. Eastman would use a Texaco gasifier to produce the synthesis gas required to make methanol and acetic anhydride. Noting that Halcon had been working in a similar area, Eastman approached Halcon and the companies decided to pool their technologies and patents.

The economics of producing acetic anhydride from coal instead of from natural gas were favorable, at least at the time the decision was made to build the plant. What helped was the fact that Eastman already had the coal receiving, handling and other infrastructure that had been built for producing steam for the entire very large Eastman chemical complex at Kingsport, Tennessee. This is important, since companies not using coal as their utility fuel, would have to install such an infrastructure to do what Eastman did (Fig. 13.1).

An important aspect of this process was the fact that the acetic anhydride was made entirely from synthesis gas, as no ethylene was required. The acetic acid used in the reaction was produced internally and recycled. This was the first commercial plant to use the Texaco coal gasification process, which is considered superior to the Lurgi process that was used by I.G. Farben when German was conducting the Second

World War. The Eastman-Halcon cooperation was successful and Eastman received the coveted *Chemical Engineering* magazine Kirkpatrick award for achievement in 1985. Halcon sold the technology to Eastman and did not design any other acetic anhydride plant. SD's engineers did work closely with Eastman to achieve successful operation of this interesting technology.

13.2   The Polyvinyl Alcohol (PVA) Problem

In some respects, this project became an apocryphal bookend to the Halcon/SD story: A third party license that a Mexican company obtained from a U.S. firm was to be used to design a plant that would serve the entire Mexican market. SD was then chosen to develop a process design, using the technology, which was based on a prewar German plant by people who hardly remembered the details. The plant was built just before the end of the 1970s.

In 1982, after Halcon had become part of Texas Eastern, there was a call from Mexico. The plant did not work! SD engineers visited the plant and found out that the plant was engineered and built by a contractor who had never seen a chemical plant before. The instruments and controls were faulty. And most importantly, the plant had been built in a different location than was originally planned, at a much lower elevation than where SD had been told the plant would be built, with much higher daytime temperatures. As a result, the vacuum equipment would not work since the cooling water was at a much higher temperature than design. The Mexican company told Halcon to fix the problem at its expense!

Halcon first declined, citing that all it originally received was a $25,000 design fee (which is usually the penalty an engineering firm has to pay for non-performance, but even that would not have been awarded in any arbitration due to the changed location of the plant.) The Mexican firm then informed Texas Eastern that if Halcon did not make the plant run, Texas Eastern would no longer be able to acquire certain raw materials it needed to buy in Mexico for another Texas Eastern business. Texas Eastern then turned to its Halcon subsidiary and said, "your problem". Neil Yeoman, who recalled this story recently, and who was then Vice President and Chief Process Engineer at SD, was quite familiar with the PVA project since he and another engineer had done the original design, but had heard nothing about the plant since the design was submitted to the client.

The story had a bittersweet ending. Neil and the second SD engineer who had helped with the original design went to Mexico to solve the problem. There were equipment changes and a number of other modifications. Neil and his assistant spent a great deal of time in Mexico. A year later, the plant, was working at full capacity making product at 86–88% yield, but Halcon had to absorb the cost: $1.3 million in 1983 dollars!

13.3 Halcon/SD and Oxiteno Look at Brazilian Alcohol as a Feedstock

With a huge potential for sugar cane-based ethanol, it was logical for Brazilian chemical companies to look at making ethylene derivatives based on this resource. Production of ethanol from cane was slated to rise from 3 billion tons in 1980 to more than 10 billion tons in 1985. A program called PROALCOOL was established by the government to provide financial and logistical assistance to companies interested in pursuing this opportunity.

Oxiteno, one of Brazil's largest chemical firms, headed by Pedro Wongtschowski, was already a client of SD's, having purchase SD's ethylene oxide and derivatives technology to build Brazil's largest plant of this kind. Halcon had, in fact, received a 10% interest in Oxiteno as a part of this transaction. Halcon and Oxiteno in 1981 signed a joint program of research and development for creating new technologies using ethanol as a feedstock. Both Oxiteno's and Halcon's laboratories would work on this program. The emphasis was not necessarily to come up with brand new processes, but rather to improve existing technologies that made "petrochemicals" from ethanol. The agreement called for Oxiteno to retain exclusive rights in Brazil and for Halcon to have exclusive rights in the rest of the world. Africa and Asia were considered promising areas for exploitation.

Little is known about this program. However, it is likely that Texas Eastern Pipeline, with vast resources of natural gas and gas liquids, was not likely to support ethanol-based petrochemicals and so the program was undoubtedly discontinued, on the Halcon side, at least.

13.4 Government Work Helped, but not Enough

The years (1982–1887) when Texas Eastern owned Halcon were disappointing. John Rehnberg, the son of the original owner was brought in to run Halcon, while Ernie Korchak ran Scientific Design. As usual, SD had to scramble for work. Aside from some additional ethylene oxide/glycol licenses and accompanying engineering design work, it was difficult to keep the staff busy. Accordingly, much effort was expended in securing funding from DOE or other government agencies.

In a new brochure issued at that time, John Rehnberg described what he termed "Developing Tomorrow's Technologies"

> The Halcon/SD portfolio of new technologies includes new carbonylation processes to manufacture acetyl compounds, including acetic anhydride and vinyl acetate monomer from synthesis gas. SD has served the U.S. Department of Energy as a technical support contractor for both coal gasification and liquefaction. SD is helping private industry evaluate the viability of rotary kiln and solvent refined coal processes and the use of coal with municipal solid waste and sewage sludge to produce synthetic fuels. In the field of cement technology, we offer a flexible effluent fluid bed process which generates electricity to supply an entire cement facility, a technology already demonstrated.

Rehnberg, Fred Gans, Neil Yeomen, Ron Cascone and other SD engineers continued to look at synthetic fuels and government spending as a means of generating needed business for SD.

Some of the projects carried out during this period are described below:

- An interesting project for NASA involved the space shuttle. It was said to have originated because of allegedly uncertain supply of natural gas to the Cape Canaveral area, but looking back it's reasonable to believe that this was another "booddoggle" project pushed for Federal support by state politicians in Florida. The work involved development of economic baseline data for competing feedstocks fed to a gasifier that would make hydrogen for the shuttle.
- A study for Wisconsin Power and Light Company involving pelletized high grade fuel production from Western coal.
- Computer modelling, using ASPEN software, of the Great Plains Coal Gasification Project.
- A project for Allis Chalmers involving processing steps to recover energy from spent shale.

By 1987, when Texas Eastern itself ran into financial difficulty, Scientific Design became a property with limited but still some value. The owners decided that there was no further point in supporting its engineering effort, except for that required to support the ethylene oxide business. It was still a viable competitor in ethylene oxide and derivatives technology and manufactured EO catalysts. Same for maleic anhydride, but there was little or no business there. SD's headquarters remained in Little Ferry, N.J. All employees not needed to support this operation were dismissed. As mentioned earlier, several companies acquired and subsequently sold SD, until it was eventually owned by Sabic and Clariant. The company remains a viable competitor and is further discussed in the last chapter.

Chapter 14
Epilogue

Abstract The rapid rise of the petrochemical industry was unique in the history of industrial innovation. The circumstances: shift to new, abundantly available raw materials, chemical engineering advancements and U.S. government investment in novel chemical plants during World War II. All were certainly propitious. And so a new industry came into being and experienced double digit growth in the 1950–1970 period. But then it ran into difficulties. And it became just another industry.

Chapters 6, 7, 8, and 9 described how Scientific Design Company and other firms developed a number of technologies during a highly inventive period. Many other domestic and foreign companies also developed new technologies or greatly improved existing processes during the same time. All of this led to the construction of more and more, larger and larger plants with rapidly decreasing production costs—a boon to consumers, but a bane to the producers in a mercilessly competitive industry unless they could somehow differentiate their offering. The common term for this is commoditization and this was the inevitable fate of the petrochemical industry.

This book has featured the role of Scientific Design Company as a leading actor in the rise of the industry. It would be a mistake to say that SD's role was paramount in the creation of the industry. It was, however, quite significant. When a group of entrepreneurs with chemical engineering skills creates a firm that can "outresearch" the large existing companies and also become the leading designer of a very broad range of chemical plants, such an achievement needs to be cherished. And that is the reason for this book.

14.1 Ralph Landau

In 1982, Landau started a second career—as a scholar. One of his main interests was to better understand how technically successful companies sustain themselves and how they fail. While at SD and Halcon he had already written and co-authored a number of articles, he then devoted the rest of his life in academic, and philanthropic

© Springer Nature Switzerland AG 2019
P. H. Spitz, *Primed for Success: The Story of Scientific Design Company*,
https://doi.org/10.1007/978-3-030-12314-7_14

Fig. 14.1 Ralph Landau in
academic retirement. *Source*
Science History Institute

pursuits. He moved to California and became a consulting professor of economics
and chemical engineering at Stanford University. There, he co-directed the Program
on Technology and Economic Growth at Stanford's Center for Economic Policy
Research. He supported Stanford and was instrumental in the construction of the
Landau building, which was completed in 1994 (Fig. 14.1).

Landau's preoccupation with academic economics was largely based on his experience with Halcon. He was preoccupied with the fact that a technically successful
firm like Halcon could not sustain itself. He examined the economics of innovation and saw himself as standing between scientists developing new technology and
academic economists and business leaders.

At Stanford, he wrote books such as *Technology and the Wealth of Nations*, which
looked at ways in which countries, government policies, and companies interact to
influence economic growth and technological advancements. He looked at technology as capital and examined the economics of innovation and globalization.

He was appointed a fellow of the faculty at Harvard's John F. Kennedy School
of Government and there co-directed the Program on Technology and Economic
Policy. Previously, he had funded much of the building cost of the Ralph Landau

Chemical Engineering building at MIT, created an endowment for the MIT Chemical Engineering Practice School, which gives students experience working in industry and was a member of the Visiting Committee at MIT.

As a distinguished alumnus, he established a challenge fund at the University of Pennsylvania and sat on the Board of Overseers for the School of Engineering and Applied Sciences at Penn.

Landau was elected to the National Academy of Engineering and as a foreign member of the Royal Academy of Engineering (United Kingdom.).

Among the many awards and honor conferred on Landau were the following:

1973 Chemical Industry Medal, Society of Chemical Industry
1977 Winthrop Sears Medal, Chemists Club
1981 Perkin Medal, Society of Chemical Industry
1982 Founders Award, American Institute of Chemical Engineers
1985 National Medal of Technology
1997 Othmer Gold Medal, Chemical Heritage Foundation
2000 Petrochemical Heritage Award
2003 Lifetime Achievement Award, Lester Center for Entrepreneurship and Innovation

Chemical Engineering Progress magazine posthumously named Landau one of "Fifty Chemical Engineers of the Foundation Age."

Ralph Landau passed away on April 5, 2004.

Ralph and Claire Landau's only child, daughter Laurie, earned an advanced degree in Equine Veterinary Medicine at the University of Pennsylvania. She has engaged in various philanthropic activities and is currently chair of the Science History Institute (formerly Chemical Heritage Foundation) in Philadelphia, Pa.

14.2 Propylene Oxide: The Legacy of a Significant Achievement

When Ralph Landau was forced to sell Halcon's share of the Oxirane joint venture to the other partner, Arco Chemical, it was the end of a dream. The unfortunate decision that caused Halcon to push ahead with the construction of the eventually doomed ethylene glycol (MEG) plant turned out to be a major mistake that Landau must have regretted for the rest of his life. A normally extremely cautious man, he went ahead with the project in spite of warnings from many of his trusted senior staff that the technology was not ready for commercialization or that the design did not have enough safeguards against unknown risk. This is hard to understand in retrospect. He was evidently in a hurry to make Halcon the large financial chemical company that could go public with all the associated benefits. That did not happen, but he and the Rehnberg family did amass a large personal fortune at the buyout.

The astounding success of the propylene oxide technology, researched separately by SD and Arco Chemical and then combined into Oxirane, is a tribute to SD and

Ralph Landau. The combined patent position of these two firms kept other firms, principally Shell Chemical, from becoming a competitor in this technology. This gave a lasting global position on propylene oxide to Lyondell Petrochemical, the operator of two large ethylene crackers and a refinery which absorbed Arco Chemical after both of these firms were spun off by Atlantic Richfield and floated to the public in the 1980s… Over the following decades, much more propylene oxide capacity was built and the scope of the enterprise was greatly expanded, including into China. Some of the expansions were carried out in joint venture with Bayer, which is a large polyurethane producer. In a recent annual report, Lyondell Basell has published its global capacities of different parts of the propylene oxide/byproduct family of products as follows: Propylene oxide—5.1 billion pounds, styrene monomer—5.9 billion pounds, and tertiary butyl alcohol—5.8 billion pounds.

This represents one of the most important parts of Ralph Landau's legacy, which must be shared with Arco Chemical, since both SD and Arco decided to form Oxirane, leading to a unique achievement.

In 2006, Dow and BASF announced a joint venture to build a 300,000 tons per year propylene oxide plant in Antwerp, Belgium, based on direct oxidation of propylene with hydrogen peroxide. The two firms had carried out research in this area and had decided to merge their technologies. Other firms, notably Degussa in Germany, had also been working on such a route and announced a 100,000 tons per year plant in South Korea. In an article published in *Chemical & Engineering News* on October of that year, Dr. Jeffrey S. Plotkin at Nexant/Chem Systems opined that both the co-product and the peroxide routes look good for the future, the PO/Styrene technology very competitive when a company wants to make both propylene oxide and styrene.

14.3   Scientific Design Company: Still Competing on Ethylene Oxide

A final version of the company that Landau, Rehnberg and Egbert created contin-ues to provide engineering services and catalyst manufacturing, but these are almost exclusively limited to ethylene oxide, glycol and glycol ethers, as far as new project announcements are concerned. It also offers process technologies on ethanolamines, ethoxylated products and polyesters, according to the company's website. The com-pany has remained at its long-time location in Little Ferry, New Jersey and has satellite offices in Bahrein and China.

The market for ethylene oxide and glycol plants remains strong, particularly in China, where, in 2016, SD licensed technology and supplied the catalyst for a 100,000 ton per year ethylene oxide plant for Sanrui Technology Chemical Company. SD continues to compete against Shell/Lummus, as well as some other, most likely' Japanese firms offering the silver catalyst direct oxidation route.

Several years ago, Shell came out with a different version of the conventional ethylene oxide/glycol technology called the *Omega* process. In this variation, the ethylene oxide produced is not converted to ethylene glycol but reacts with carbon dioxide to yield ethylene carbonate, which is subsequently hydrolyzed to monoethylene glycol. No di- or –triglycols are formed in this process. Mitsubishi Chemical Industries invented this technology and licensed it exclusively to Shell.

According to Shell, the selectivity of EO to MEG for the Omega process is 99.3–99.5% and it can produce up to 1.95 tons of MEG from 1 ton of ethylene. The conventional process offered by Shell produces 1.53–1.70 tons of MEG per ton of ethylene, with SD probably achieving a similar result.

The Omega process was first implemented in South Korea and subsequently in Saudi Arabia. Shell in 2010 announced a 100,000 ton per year MEG plant it was building in Singapore.

To date, no company has developed the theoretically most efficient process for making ethylene glycol: the acetoxylation route, which brought about the demise of Halcon.

14.4   The Petrochemical Industry

The two oil shocks of the 1970s had a major effect on the global economy and also on the U.S. petrochemical industry. The world crude oil price increased from three dollars a barrel to over thirty dollars, making petrochemical feedstocks like naphtha and gas oil much more expensive when produced from crude oil fractions. Over the same period, U.S. natural gas prices also increased dramatically from 30 cents per million BTU to two dollars or higher by 1980, as industrial users and utility companies shifted from petroleum fuels to natural gas and suppliers of gas could barely keep up with demand. There was a brief period when petrochemicals producers made "windfall" profits selling products made from earlier cheap fuels into a temporarily high priced market, but higher production costs and corresponding lack of international competitiveness brought petrochemical operating rates down substantially and the industry went into recession. By 1982, petrochemical producers did all they could to cut costs, including research, and stopped all new capacity additions. Demand had already slowed as the substitution of synthetics for natural materials had pretty much run its course. So, the industry hunkered down, waiting for better days.

But some experienced investors knew that demand would again catch up with supply and maybe sooner than many realized. Gordon Cain, an ex chemical executive approached DuPont and asked whether the company would sell its Conoco Chemicals subsidiary, acquired when DuPont several years earlier had bought Continental Oil Company. After months of difficult negotiations, Cain's group made the first chemical leveraged buyout, creating Vista Chemical. Other such transactions followed, including a huge second buyout by the Cain Group. By 1987, petrochemical operating rates had moved up appreciably and by 1988 prices hit an all-time high. In the space of only a decade, two pricing cycles occurred! As industry participants predictably reinvested in new capacity, operating rates dropped sharply again in a couple of years, then reached another high in 1995 as a major hurricane and other plant outages cut capacity. It was now time for participants to reassess their interest in staying in an industry where year-to-year profits could sharply change. By the end of the century, the industry had started to fall into a pattern.

- All large U.S. chemical companies except Dow soon decided to leave the petrochemical industry.
- Several of the large oil companies (Exxon, Shell, Conoco-Phillips) stayed in, citing their strong back integration and position as global petrochemical producers.
- Some private companies (Ineos, Huntsman) had acquired petrochemical assets at low prices and found a reason to continue.
- Several Asian producers (Shinetsu, Formosa Plastics, Westlake) also decided that U.S. petrochemicals were a reasonable place to be.

   With the advent of the shale era at the end of the first decade of the next century, these more steadfast participants were rewarded, as domestic prices for natural gas and gas liquids plummeted, providing again the competitive advantage that U.S. producers had originally enjoyed with cheap natural gas on the Gulf Coast and in the Appalachia region. U.S. producers became as competitive as Middle East companies until things equilibrated and that is where we are at this writing.

   Reverting to technology development, it is fair to say that the petrochemical industry was never as inventive again as it was in its early years. The efficiency of processes increased steadily, with more selective catalysts. Some new technologies were commercialized, but there were no major breakthroughs and no more billion pound polymers. Probably, the most important new technology was the production of ethylene from synthesis gas, the so-called MTO (methanol-to-olefins) technology developed by Mobil and UOP, now used principally in China, where coal is the dominant local resource. Much effort was expended to develop processes for converting ethane or propane directly to petrochemicals (known as alkane activation) but this has not yet been achieved in any important commercial sense. It is also fair to say that much of the new research in petrochemicals is now carried out in China, Japan and in the Middle East.

   In China, ethylene and propylene derivatives are made in large quantities from coal using the above-mentioned MTO technology. These plants have very high capital investment, but relatively competitive operating costs. This technology is well suited to China's raw materials situation and government incentives, but not applicable for other parts of the world. The only exception is Sasol in South Africa, which continues to produce fuels and chemicals using Fischer-Tropsch technology based on coal.

   It is reasonable to speculate that for the foreseeable future, petrochemicals will be made from hydrocarbons. With automobiles will increasingly powered by electricity, there will be little concern about the world running out of crude oil, with petrochemicals increasing its share of the barrel.

14.5   Chemical Engineers Turn to Other Challenges

The early years of the petrochemical industry, were a time when chemical engineers came into their own. What chemical engineering students learned in class, they were able to apply in industry and so a large number of chemical processes were developed and commercialized. But as the petrochemical industry matured, the opportunities for chemical engineers in heavy industry started to decline.

An important theme of this book is the fact that, in important ways, chemical engineers created the petrochemical industry. Why chemical engineers? Because this engineering discipline could take a laboratory invention that resulted in a few grams of a new material produced from millions of dollars of research and make it available by developing a process and designing a manufacturing plant that could produce the material at a typical price of 20 cents per pound. This was done for all of the important petrochemicals, liquid and polymer alike, until no more such opportunities remained.

If the petrochemical industry no longer offered chemical engineers an area for breakthrough inventions, could they find similar opportunities in another area? As it turned out, the timing was right for such a consideration. In the aftermath of the two oil shocks and the resulting high crude oil prices, as well as increasing concern in some quarters about using up valuable resources, the idea of making "petrochemicals" from renewable materials started to be of interest. A few large scale organic chemicals, notably ethyl alcohol and citric acid, had long been produced using fermentation. The fact that, in Brazil, sugarcane-based ethanol was very inexpensive and would soon be producing ethylene and polyethylene was proof positive that bio-based technology should be pursued. Additional impetus for a bio-based approach was the concern that hydrocarbon-based plastics are not biodegradable in landfill or elsewhere, creating a desire to determine whether similar plastics could be made from renewable materials, such as starch. In fact Cargill and Dow were producing a plastic product of this type, while firms like Coca Cola and Pepsi were encouraging the chemical industry to produce the ingredients for polyester bottles from biomass.

Efforts along these lines occupied chemists and chemical engineers over the 1980–2000 period, as crude oil prices and to some extent natural gas prices remained high much of the time. Bioengineering became a "hot" area and added a new frontier in chemical engineering. Many U.S. chemical engineering departments were renamed "Chemical and Biomolecular Engineering Departments or "Chemical and Biological Engineering Departments".

There was some progress in visualizing and, to a limited extent, developing biological processes that could compete economically with existing petrochemical routes. This included the idea to gasify biomass, thus producing carbon monoxide and hydrogen that could be further processed to organic chemicals and polymers. However, another issue would cause developers to wonder whether a biomass approach was feasible. This involved not the scale-up of the technology, but that of collecting and transporting the vast amounts of biomass (corn stalks, grasses) that would be required to feed a world-scale plant making a product like ethylene glycol or ammonia. Becoming increasingly carbon emission conscious, developers also worried about the large amounts of carbon dioxide that would be given off by the trucks collecting and transporting the biomass.

But the most important reason why this area of development was essentially discontinued was the shale oil "revolution" that occurred early in this century. Natural gas prices and that of the associated ethane and higher aliphatics plunged to levels not seen since the heyday of the petrochemical industry and so the impetus to develop large scale biomass-based technology quickly disappeared.

A paper published in 2017 by Ignacio Grossman of Carnegie Mellon University noted that Dow Chemical had become concerned about "the big push to Bio", stating that it needed to hire 75–100 Ph.D's per year having skills not only in classical chemical engineering, but also in such areas as material sciences, energy storage, lightweight materials and electronic materials. It also stated that in universities there is decreased emphasis on chemical engineering fundamentals, transport courses, thermodynamics, phase and equilibria), with process design courses in some department outsourced to retired industry people (!). Moreover, a 1913 industrial survey of Chemengineering recruiters and leaders regarding desired skills of relative importance to companies was enlightening. It put unit operations, thermodynamics and separation processes at highest level of importance and biotechnology and nanotechnology at lowest.

In October 2016 the MIT School of Chemical Engineering Practice celebrated its 100th anniversary. Professor Alan Hatton, the current director, said some words that seem fitting to close this section.

Life has changed a lot since 1916, but the core of the Practice School experience for student s and hosts hasn't. For a century, chemical engineering students have worked with industry partners to help solve real problems in pharmaceutical, energy, food sciences, materials, the biomedical field and other areas.

The venues have changed- back then they were confined to the Northeastern U.S., while today students travel to stations around the world—but the fundamental tenets of the Practice School have remained the same. The Practice School has instilled in generations of students critical thinking and analytical skills, tenacity, cultural sensitivity, confidence and a desire to do right by the company that has opened its doors to them. Whether students are working with carbon black at Cabot, reengineering Golden Graham cereal at General Mills or analyzing the surface qualities of an antacid pill at Merck, they are contributing to the success of their team and host company, while learning skills one can only learn through such a dynamic and intensive experience.

Times have changed, but the mission remains the same: To immerse students of chemical engineering in industry, to present them tough but real problems that require solutions, and to challenge them to use their training, their wits, their teammates, and the industrial experts around them to shine light on those problems to the benefit of all.

Many of the chemical engineers that worked at Scientific Design Company went through the MIT Practice School. It seems reasonable to give thanks to and celebrate the founders of the Practice School for enabling a group of entrepreneuring chemical engineers to achieve what they did at this unusual company.

14.6   The Unusual History of Petrochemicals

This book has recounted the rapid rise of the petrochemical industry that took place over three decades from the end of the Second World War to the late 1960s or 1970s. By that time, coal-based and alcohol-based chemicals had almost completely been displaced by petrochemicals, while synthetic fibers production greatly exceeded that of natural and manmade fibers. This was not only a case of raw materials substitution, but also the invention of brand new polymers (polyethylene, PVC, nylon, polyester) barely known before the war. This raises an interesting question: How unique was this industrial experience? What other important industries underwent such a dramatic disruption over such a short period of time? Is there a historical parallel to a situation where an industry went through such a dramatic change with an entrepreneur at its head? It was important to gain this historical perspective.

The petrochemical era has two characteristics. There was a complete change in the raw materials from which a number of important chemicals were produced. And there were also a number of new products (thermoplastics, synthetic fibers) that greatly added to the volume of materials produced by the industry. Conversely, there was little equipment innovation, as petrochemicals production used much of the same type of equipment already in use in petroleum refineries and ammonia plants (distillation columns, reactors, heat exchangers pumps, compressors) and in the production of inorganic and organic chemicals (evaporators, crystallizers, filters).

Synthetic dyestuffs: This may be the closest analogy. Until synthetic dyes were discovered, natural dyes were derived from plants and, to a lesser extent, from shellfish and insects. The rapid growth of the British textile industry created a need to manufacture larger and cheaper quantities of dyestuffs. In 1956, William Henry Perkin, in trying to synthesize the anti-malarial drug quinine, using a coal tar chemical related to aniline, serendipitously discovered a purple material which in alcohol could readily dye silk. He immediately started to manufacture *Perkin mauve* on an industrial scale. Other aniline dyes were discovered over the next five to ten years. In 1869, alizarin, a blue dye based on anthracene was discovered at BASF. Azo dyes were also discovered around 1975. By 1877, Germany accounted for more than half of the world's manufactured synthetic dyes. No useful statistics were found, but unquestionably synthetic dyestuffs by the latter part of the nineteenth century had, by volume, greatly surpassed the production and use of natural dyes and had become the most valuable large scale chemical in the world. There is an apt comparison to petrochemicals in terms of (a) substitution of synthetic for natural materials, (b) new, large volume products (many new, different color dyes) and (c) a period of only 20–30 years when the new technologies took over a large part of the annual production volume of all dyestuffs.

A quick look at a number of other industries uncovered no other cases that would qualify. The following were considered:

- *Iron and Steelmaking* went through a number of phases from the Bessemer process in 1857 to open hearth steelmaking and the rise of integrated steel mills over the next fifty years or so, followed by the basic oxygen furnace. A long term transformation.
- *Aluminum* was first produced electrolytically in 1854 in France and Germany. Large scale production using molten cryolite, which reduced the melting point of aluminum, occurred in 1888 in France and in the United States. This method was shortly thereafter improved by purifying the bauxite. Then, in 1920, the use of continuous electrodes was invented. Again, a long period from invention to maturity,
- *Papermaking* became an industry in the early 19th century in Europe with the invention of the Fourdrinier machine which produces a continuous sheet of paper rather than individual sheets. It was initially run on (cellulose containing) waste rags. Then, in 1844, two inventors started to make paper from wood pulp and the process has remained essentially the same since then.
- *Sheet glass* was first produced in England in the mid-19th century, using a cast iron bed and rollers. The current, modern technique, known as the float glass process, was invented by Pilkington, again in England, in the 1950s. This involved using a continuous ribbon of glass formed on a molten tin bath, giving the sheet uniform thickness and flat surfaces.

Manufacturing industries advance as new technologies are invented. This process tends to be slow in most cases. But there are times where a disruptive change can occur and this may come about when circumstances make this possible. In the case of dyestuffs, the industrial revolution in Europe, with the rapid growth of the textile industry, created a need for more diverse dyes and in greater amounts than suppliers could fill. At the same time, research at that time on coal tar derivatives like aniline, anthracene and phthalic anhydride, including a much better understanding of these molecules and their derivatives, rapidly led to the discovery of a number of synthetic dyes. The new industry was created because circumstances were propitious. The same was true of the petrochemical industry, but in reverse order. The chemistry required to manufacture both old and new products from hydrocarbons was ready for rapid commercialization. And then the need for these products by the war effort and, not much later, by consumers with money to spend and long-delayed investments in housing, cars and other goods spurred a somewhat lethargic chemical industry and a number of new entrants to build a new industry to fulfill that need. Since the supply of coal- and alcohol-based chemicals was limited, hydrocarbon feedstocks took over. For people who have been involved in chemistry in their professional life, it is gratifying that this area of knowledge was there and ready to serve when the populace needed it. Who would have thought that synthetic dyestuffs could so quickly supplant and surpass the many natural dyes used to that time? And who in

1936 could have predicted that twenty years later an unknown product like polyester fiber would be the go-to fabric for carpets and textiles.

So, the rise and growth of the petrochemical industry was relatively unique in the annals of industrial development. My first book, *Petrochemicals. The rise of an industry* traced the story in a true historical perspective, describing the development and international growth of the industry in terms of a series of events and examples that allow the reader to understand the origins and significance of a disruptive change in an established industry. The current book uses the foundation and growth of Scientific Design Company and the participation of many talented chemical engineers at SD and in other firms to personalize the creation of the petrochemical industry, such an unusual and fascinating happening.

Correction to: Primed for Success: The Story of Scientific Design Company

Correction to:
P. H. Spitz, *Primed for Success: The Story of Scientific*
Design Company, **https://doi.org/10.1007/978-3-030-12314-7**

The original version of the book was revised. The corrections which were inadvertently omitted in the original version have now been included. The erratum book has been updated with the changes.

The updated version of the book can be found at
https://doi.org/10.1007/978-3-030-12314-7